On Your Own

How to Escape the Corporation and Make More Money as an Independent Contractor

Carol S. Lewis
AND
Harry L. Helms

HighText publications inc.
Solana Beach, California

Library of Congress Cataloging-in-Publication Data

Lewis, Carol S. (Carol Surber), 1949–
On your own : how to escape the corporation and make more money as an independent contractor / Carol S. Lewis, Harry L. Helms.
p. cm.
Includes index.
ISBN 1-878707-13-2 (paper) : $14.95
1. Independent contractors. 2. Self-employed. 3. New business enterprises—Management. I. Helms, Harry L. II. Title.
HD2365.L49 1994
658'.041—dc20 94-4847
CIP

ISBN: 1-878707-13-2
Library of Congress catalog number : 94-4847

Printed in the United States of America
Cover design: Brian McMurdo, Ventana Studio, Valley Center, CA
Interior design and production services: Sara Patton, Wailuku, HI

P.O. Box 1489 ✦ Solana Beach, CA, 92075

Contents

Chapter 8

Random Thoughts 185

Introduction

Independent contracting turns the traditional employer/employee relationship on its head. Instead of having a single employer who tells you what to do, when and how you will do it, and how much you will be paid, you have multiple "employers." The terms and conditions of the work you perform—such as precisely what you will do and what you will be paid for doing it—are not dictated to you. Instead, they're determined on a case-by-case basis through negotiation with the individual, company, or organization paying you.

Independent contracting is *not* the same as temporary employment (although the popular media often confuses the two). It is *not* something you do between permanent jobs, or on a part-time basis, or until you figure out what you really want to do with your life. Independent contracting is a legitimate career path in its own right.

You might think that a full-time employee of a large company or organization has more "job security" than an independent contractor. Actually, the opposite is true. People who are full-time employees have only one customer for their services—their employer—and usually have no idea how heavily dependent they are on their employer for financial support and self-esteem. If these people lose their job, they are often in deep trouble because they have completely lost the skills essential to professional and personal self-sufficiency. By contrast, independent contractors work for multiple organizations and are not heavily dependent on any one of them. Independent contractors know they will lose a certain amount of existing business each year, but also know how to gain new business to replace the losses. Independent contractors are psychologically autonomous and self-reliant. They don't wait for good things to happen in their career. Instead, they go out and make good things happen.

Independent contractors usually earn more than full-time employees performing comparable tasks within a company or other organization. As we'll discuss later in this book, independent contractors sometimes attract more business by *raising* the prices of their services!

When you become an independent contractor, you have much greater control over your professional fate. That sounds nice—and it is—but it also comes with a heavy load of responsibility. Independent contracting isn't for everyone. If you lack self-discipline, if you're uncomfortable taking the initiative in a business situation, if you dislike having to sell yourself and your skills to strangers, or if you want someone else to tell you how much you're going to be paid for doing something, then independent contracting isn't for you.

Independent contracting isn't restricted to technical specialists like computer programmers or accountants. Almost anyone with a few years of experience in a certain field will have skills that individuals, companies, and organizations will pay for on a per-task contract basis. The same is also true if you have knowledge of a new, emerging field. The recent trend of corporations of all sizes to "outsource" many functions has meant that skills as diverse as warehouse management, employee training, marketing, product design, government contract administration, human resource management, corporate communications, quality control, purchasing, and engineering are bought on a contract basis. Independent contracting has also been fueled by the growth of small businesses which need specialized expertise and skills occasionally, but not frequently enough to justify a full-time employee for that function. For example, the cover of this book was designed by an independent contractor and our corporate accounting is likewise handled on a contract basis with an accountant who handles accounts for several small firms.

There is no magic formula for succeeding as an independent contractor, and the methods we'll describe in this book are not the only possible ways to go about contracting. Other successful independent contractors might disagree strongly with what we advocate in these pages. However, these are the techniques that work for us and many other contractors we know. Our contracting activities involve technical manual and multimedia software preparation for the electronics and computer industries, but the same principles we'll describe in the pages ahead apply no matter what service you offer on a contract basis. At HighText, we also make extensive use of independent contractors for such things as technical artwork and page layout, and thus also know how independent contracting works from the buyer's perspective. This book was not brought down from Mt. Sinai and the advice we give won't apply to every contractor in every field; filter what we say through your experience and decide how valid it is for your situation. However, we feel the broad principles we discuss—particularly how to find potential clients, conduct meetings with them, and get agreements in writing—will apply no matter what you do on a contract basis.

By the way, this book won't cover such topics as bookkeeping, taxes, legal requirements, and other topics that are important to anyone working as an independent contractor. There are plenty of good books and materials describing those subjects, and we can't claim any expertise in such areas. Instead, this book focuses on the *process* of independent contracting.

Independent contracting is more than a temporary response to the chaotic economic environment of the 1990s. It is a course of action that allows you to put your priorities and desires first—instead of an employer's—in your life and career. It is not always easy, nor is it always fun; many people find independent contracting is more difficult in many ways than working as a conventional employee for a company. But others find independent contracting more rewarding and liberating than any conventional job can ever be. Indeed, a common observation by successful independent contractors is "I could never go back to working for a company again!"

Want to make that statement yourself? Start reading . . .

Carol S. Lewis ✦ Harry L. Helms
co-owners, HighText Publications, Inc.

Chapter 1

The Alternative To Corporate Employment—Independent Contracting

Independent contracting is sometimes portrayed in the media as some sort of wacky fad—like the disco craze of the 1970s—that will fade away once the economy improves and Fortune 500 companies start hiring large numbers of permanent employees again.

But independent contracting isn't a transitory phenomenon. It will be a long time—if ever—before most major companies have as many permanent employees as they did in the 1970s and 1980s. If you're looking for career stability and income growth in the future, don't go looking in Fortune 500 companies. You'll have better odds for stability and money as an independent contractor.

Independent contracting isn't the same thing as temporary employment, even though the media often confuse the two. Temporary employment is almost the exact opposite of true independent contracting. Temporary employment is nothing but a variation of the old employer/employee game, except that temporary employees get no benefits (such as medical coverage or a retirement plan), have zero job security, and have no real assurances about the future other than he or she will be out of a job again soon.

You might look at the large-scale layoffs and staff cutbacks at major corporations and conclude that such companies don't need independent contractors either—if they did, so the reasoning goes, they wouldn't be cutting permanent employees. But the truth of the matter is that the same economic forces that cause companies to trim the number of full-time employees increase opportunities for independent contractors. In fact, it's not too much of an exaggeration to say that what's bad for full-time employees is good for independent contractors. A powerful set of economic, social, and technical forces have been unleashed in the United States, and these forces will continue to radically redefine how companies *must* operate to survive and prosper. Independent contractors will be an increasingly important part of any company's efforts to cope with such changes.

Why Companies Need Independent Contractors

So why are companies using more independent contractors at the same time they are cutting back on the number of full-time permanent employees? Indeed, a lot of corporate managers would probably prefer to keep and use permanent employees instead of relying on independent contractors for important tasks. (Having a large number of employees reporting to you is one measure of corporate macho, after all!) But there are several powerful reasons why corporations use independent contractors instead of employees. Here are some:

✦ **Certain tasks must be performed for a corporation, yet those tasks aren't done frequently enough to justify a permanent employee.** Accountants and lawyers are two traditional examples of those who perform tasks that every company needs from time to time, yet many smaller companies do not require such services often enough to justify a full-time employee. Smaller and start-up companies rely heavily on independent specialists in such fields as computer programming, marketing, public relations, desktop publishing, engineering, graphic arts and design, and prototype construction. If the need for a contractor's services is regular enough (as in marketing or public relations tasks), some sort of retainer or yearly contract relationship is often possible. Other tasks are only needed by a company rarely (such as translation of a company brochure into a foreign language) or on a one-time only basis (like the design of a corporate logo). Most new permanent jobs in recent years have been created by small businesses, and small businesses are also among the biggest users of independent contractors.

✦ **Companies want staffing flexibility.** For many reasons, companies are reluctant to add permanent employees for projects that are speculative (such as a new product that might fail in the market) or for upturns in business that might be temporary. It is much easier for a company to use contractors to fill in the gaps until a new product is proven or the new level of business seems permanent. Ending a relationship with an independent contractor is much easier (legally, financially, and emotionally) than terminating a permanent employee and has far less negative impact on company morale.

✦ **Some newer companies are set up to use independent contractors instead of permanent employees.** You've probably read about "virtual corporations," which are usually defined as groups of independent contractors assembled by a company for a specific project or function. We use the "virtual corporation" concept at HighText Publications. Look at the copyright page of this book; you'll see that the cover design and page layout were done by independent contractors. This book was sold to the store you bought it from (or the library you borrowed it from) by independent sales representatives that sell books by several noncompeting publishers. Our warehousing and order fulfillment is also handled by an outside organization. Other tasks, like accounting and public relations, are also performed by independent contractors. This sort of arrangement lets HighText select the mix of talents and skills that's best for a particular book and is a great source of new ideas and concepts. Virtual corporations tend to be found in companies on the cutting edge of technological innovation; for example,

HighText works with books that are written, edited, and produced exclusively using computers instead of conventional "cut and paste" editing and production methods. If you have up-to-date skills in a high-technology area—such as electronics, computer software, or multimedia development—you will probably find abundant opportunities in new companies using the virtual corporation model. You can also sell other business services such as accounting or financial management to such virtual corporations.

✦ **Companies need special skills for special projects.** Companies, particularly in technical fields, often need specialized skills for narrowly defined, nonrecurring situations. For example, when the first Apple computer was being designed, the designer, Steve Wozniak, had no trouble with the main computer board. However, he lacked the necessary background to design an adequate power supply. That task was given to an independent contracting engineer. (The contracting engineer's fee was a few thousand dollars. He was offered some Apple stock instead. He declined, insisting on the cash. Had he taken the stock, he would have been a multimillionaire when Apple went public six years later!)

✦ **Many permanent employees laid off by companies have been in positions made obsolete by technology or that were unrelated to a company's main business.** Every major advance in technology has resulted in job losses in certain occupations, even to the point where entire job categories (coal tenders on steam locomotives, landline telegraph operators, or vacuum tube electronics engineers, for example) were completely wiped out. This process has accelerated in recent years. Computer technology has made it possible for companies to "automate" many white-collar functions and use fewer people to get the same results. For example, one employee with a personal computer can process as many forms and documents as three or four people could manually. Personal computers have even eliminated many positions for long-time employees whose skills lie in mainframe or minicomputer systems and languages; entire "data processing" departments at some companies have been replaced by networks of personal computers using off-the-shelf software. Many other people who have lost their jobs in recent years have been—to put it accurately if impolitely—corporate fat and deadwood. There have been too many senior assistants to associate divisional vice-presidents running loose in corporate America, all sending memos to each other and eating away at profits. While many such people did work hard at their jobs, they produced nothing that customers were willing to pay for or that was essential to corporate functioning. None of this is meant to minimize the pain of workers who lose their jobs through technology changes or corporate reorganizations. However, such job losses do not accurately reflect the need of companies for people with contemporary skills in areas relevant to the company's primary business areas.

✦ **Government regulations and civil litigation are making permanent employees more of a burden than an asset.** Many well-meaning measures intended to benefit permanent employees have instead made them more expensive and burdensome to employers. Such mandated benefits as family leave time (even if unpaid) and health insurance are not free and must be paid for by employers. Lawsuits by terminated employees are common, regardless of how well-founded the reasons for termination. To protect themselves, many employers resort to "defensive personnel management" practices designed to compile air-tight defenses by excessively documenting any disciplinary action taken against an employee. Suits for sexual harassment or various forms of racial, sexual, or age discrimination are also increasingly common, regardless of whether the grievance is legitimate or not. And some states seem to go out of their way to provide disincentives to hire permanent employees. For example, California's system of worker compensation provides generous benefits for "work stress" disabilities and requires a low burden of proof for persons making such claims. As a result, worker compensation insurance premiums for California employers have skyrocketed, greatly increasing the true cost of a permanent employee to a company. Similar disincentives to hiring permanent employees exist in other states.

✦ **Managers need help with a project but don't want to add to their "managerial overhead."** There is so much time available to a manager, and adding a new employee, even on a temporary basis, means some of that time has to be devoted to that employee. By contrast, independent contractors work—just like the term implies—independently. By using independent contractors, managers can add resources quickly without adding excessively to their supervisory duties.

✦ **Managers often want to circumvent existing corporate structures and procedures.** If something must be done quickly or a radically new approach is needed, it is typically easier and more effective for a manager to bring in an independent contractor instead of trying to fight corporate bureaucracy or change the company's culture. Contractors exist outside the corporate structure and work independently of it, and are thus not constrained by that structure.

✦ **Companies need a new perspective or insight on a situation.** Because they look at things with a fresh, unjaded pair of eyes and have no emotional attachments to previous decisions, independent contractors can recognize things about a situation that permanent employees can't because they are too close to it. Wise managers understand that an independent contractor can be good insurance against mistakes due to "group think."

✦ **Companies need a new skill or approach immediately.** It takes time to locate, hire, and bring a new permanent employee along until they are able to make a full contribution to the company. By contrast, one of the defining features of an independent contractor is the ability to hit the ground running and immediately start making positive contributions.

✦ **Companies need an outside player for an internal corporate political drama.** Sometimes a contractor is brought in for reasons that have little or nothing to do with solving a problem. In one of our assignments, the manager that retained us made it clear—without explicitly saying so—that our task was to make certain criticisms of the company's owner's manuals and suggest changes to them. Because of the internal political relationships at the company, the manager hiring us could not make such criticisms himself. If you make independent contracting a career, you'll eventually encounter situations where you are being hired by a manager who wants to show that he or she is doing "something" about a problem ("I've got it under control—I've retained an outside expert to handle it."). Sometimes you'll be hired by managers who aren't sure exactly what, if anything, they expect you to do, or are just looking for a scapegoat they can dump the blame for a problem on. (We'll discuss how to handle these situations later.)

The Big Rip-Off: Temporary Employment

The key to understanding what independent contracting is all about is the word "independent." Nowhere is this better illustrated than in the difference between true independent contracting and temporary employment. Temporary employment is growing rapidly; in 1993, *Time* magazine ran a cover story called "The Temping of America." Temporary employment is often referred to as "contract employment" or in similar terms in the media or by the agencies that supply temporary workers to companies, so it's not surprising that many people think that temporary employment is independent contracting. But, as we noted earlier, temporary employment is nothing more than the equivalent of a full-time job with little, if any, of the benefits and legal protections afforded workers directly employed by a company.

Temporary employment was once restricted to general day laborers, farm workers, and other low-skill positions. Today, professional and technical employees such as computer programmers, engineers, chemists, industrial designers, accountants, and even corporate managers are hired by companies on a temporary basis. This change is reflected in the names temporary employment agencies use. Years ago, such agencies were upfront about the temporary nature of the work they offered by including such phrases as "temporary services" in their corporate names. Today, they often use corporate names similar to those used by permanent employment agencies and seldom make reference to the temporary nature of the work they offer. A few are even using names like "xxxx Contract Professional Services" to further hide their true nature.

The key feature of any so-called "contract" temporary employment is the presence of a temporary employment agency—often referred to as a *job shop*—between the worker and the hiring company. Workers do not work for the company using their services; instead, workers are employed and paid by the job shop. The company's relationship is directly with the job shop, and the company contracts with the job shop for so many workers possessing desired skills for a period of time.

While there are scattered exceptions, employment through a job shop is very close to working for a white-collar sweatshop. When working through a job shop, *you have all of the responsibilities and problems of a permanent full-time position with none of the benefits.* The company specifies the work you will do and the schedule you will follow, supplies necessary tools (computers, copying machines, laboratory instruments, etc.), and you work under the direction of the company's managers. You are often paid less than an equivalent permanent employee for the same work because the job shop makes its money by taking a cut of the amount paid by the company for your services. You receive no employee benefits such as health and life insurance, pension, vacation, or relocation assistance. If the assignment is at a company outside your normal commuting area—on the other side of the country, for example—you may be responsible for paying your own travel and temporary housing expenses. Your work may be terminated at any time without notice, and in some states you do not have the same unemployment insurance protection that permanent full-time employees have. The work you perform and your working conditions (hours you are expected to be on the job, etc.) can be changed by the company with little recourse open to you—you are working for the job shop, after all, not the company.

Sounds like a great life, huh?

So why do temporary employees put up with this sort of treatment? Unlike independent contractors, temporary employees haven't been able to break free of the dependency bond—much like the one between a parent and child—that forms between permanent employees and employers. When you're a permanent employee, you leave a lot of parameters of your employment—what you will do, where you will do it, how you will do it, how much you will get paid, when you must be at work, etc.—up to your company and your boss. Temporary employees are often former permanent employees who still have a strong need for a "boss" to tell them when, where, and how to use their skills and how much they should be paid for using those skills. They need someone else to identify tasks that need to be performed within an organization and to designate them as the one to perform those tasks. For such people, a job is something other people give them, not something they create for themselves.

Independent contractors are different. All successful independent contractors have one trait in common: *they don't need someone else to tell them how to use their skills in order to make money.* Unlike temporary employees, independent contractors are psychologically emancipated from the traditional employer/employee mindset.

This goes to the heart of what it means to be an independent contractor. If you need someone to be your "boss," then independent contracting isn't for you.

Do Companies Prefer Working With Job Shops Instead Of Individuals?

One way some job shops try to justify their existence is by claiming that companies prefer to hire temporary employees through an agency rather than directly contracting with individuals. Is this true?

Some managers in larger companies *do* prefer to work through job shops, especially if they need several employees with broad, nonspecific skills on a temporary basis for a project. It's much easier for such a manager to tell a job shop "I need a dozen C computer language programmers" or "Send over five electronics engineers with digital experience" than it is to locate and interview a dozen C computer language programmers or five electronics engineers on an individual basis. This is especially true if the skills required in such cases are "generic" rather than highly focused.

However, a large and growing number of companies refuse to deal with job shops at all. Too many job shops will try to foist any warm body off on companies, and those companies are becoming increasingly dissatisfied with the results they get from using job shop temporaries. We've had such experiences at HighText when trying to find technical writers and illustrators skilled in using personal computers and popular word processing and drawing software to help us with several consumer-level instruction manuals. Most of the people sent to us for interviews were clearly unqualified for the tasks we had described to the job shop, and we resented having to spend our time to find that out (that is supposedly one of the tasks a job shop performs). We had specifically told the job shops we contacted that we needed people with a heavy background in writing for end users and consumers, and that we wanted no one whose sole experience was working on so-called "mil spec" manuals for military weapons systems (such people are very common in the San Diego area). However, we found that most technical writers and illustrators referred to us by the job shops we contacted had *only* "mil spec" experience. We eventually gave up trying to locate anyone through job shops and were lucky enough to be contacted by some independent contractors who fit our requirements. And our company's experience is not unique.

What's The Difference Between Independent Contracting And Consulting?

To be honest, not much. However, consultants have earned a reputation in many companies as the sort of people who come into the firm, poke around, ask a lot of questions, write a lengthy report loaded with semi-practical suggestions for the company's president, collect a hefty fee, and then leave without actually solving anything. That's not fair to the vast majority of consultants, but it is how consultants are often perceived by mid-level managers and employees.

Calling yourself an independent contractor instead of a consultant is a useful bit of "preventive semantics." Moreover, the sort of independent contracting we'll discuss in this book involves the development and implementation of workable solutions to problems. Giving advice will be a normal part of the services you offer a company, but so will be actively assisting the company in putting those suggestions into action. Good consultants do the same thing, but if you call yourself an independent contractor you avoid the negative connotations that can result from calling yourself a consultant.

Isn't "Independent Contractor" Just A Fancy Term For "Freelancer"?

Successful freelance employees have much in common with successful independent contractors. A lot of successful freelancers are actually independent contractors (as defined in this book) while some people who call themselves independent contractors really operate more like freelancers. The tax and labor regulation agencies of the various states and federal government recognizes no difference between the two. So are there really any differences?

One difference revolves around the term "contractor." *A true independent contractor always has a written contract for the work to be performed with the organization using his or her services.* Many freelancers work off "handshake" arrangements. By contrast, independent contractors always work from a written contract. The contract covers all elements of the work to be performed and defines the responsibilities of the contractor and hiring organization. The contractor is usually the one who prepares the contract. The contract is usually not a multipage, complex document prepared by an attorney; instead, it can be a simple letter or work proposal that someone at the hiring organization signs to indicate acceptance of its terms and conditions. While simple, such contracts are fully enforceable and are a big help in preventing misunderstandings and confusion.

Another difference is how independent contractors are "proactive" when looking for assignments. Freelancers typically perform narrowly defined tasks that the organization feels are necessary. On the other hand, independent contractors are constantly analyzing client organizations and actively suggesting tasks they could perform for those organizations. Often an organization that engages the services of an independent contractor will have no clear idea of what problems it faces and how those problems should be solved. Independent contractors are often given *carte blanche* by the hiring organization to uncover problems within the organization and do whatever is necessary to correct them.

A final difference is that many freelancers charge by the hour for their services. Hourly billing is the norm in some professions (such as law and accounting), and is widespread among graphics artists, technical writers, business consultants, and others. There's no reason why any service cannot be sold to clients on an hourly basis, but many independent contractors instead charge one all-inclusive fee for each task or project completed. We'll discuss pricing strategies in detail later, but for now we'll just note that the time expended on a project has no relationship to the value a customer perceives for that project. Straight hourly pricing can be a nice way to shortchange yourself!

An Independent Contractor In Action, Or How To Get A Pool Installed

You've probably used the services of independent contractors in the past, perhaps in the areas of home construction, improvements, and repairs (people in such fields are often called "contractors"). Let's take a very common example of the sort of independent contractor we're discussing in this book: a person who wants to install a swimming pool in your backyard.

Suppose you own a house in a warm climate and have a yard large enough to hold a swimming pool. You might decide that a swimming pool in your backyard would be a good idea. You could look in the yellow pages for people who install pools or ask your neighbors who installed their pools. However, often the idea of getting a pool installed will originate with an independent contractor (strictly for the sake of simplicity, we'll refer to this contractor as "she" in the rest of this section). You might see an ad in your local paper from her, get a flyer in the mail from her, or receive a telephone call asking if you'd like more information about getting a swimming pool for your house. Regardless of where the idea originates, the swimming pool contractor does something—an ad, a mailing, a telephone call, or networking through previous customers—to make you aware of her and the services she offers.

An initial meeting is set up, usually at your home so the contractor can see where the pool is to be installed. When you and the contractor meet, you're both seeking information about each other. You want to know about the contractor's expertise, the types of pools she installs, what guarantees she offers, and the names of any satisfied customers she can give as references. The contractor will ask questions to determine just what kind of pool you have in mind. You'll be quizzed on such things as the general shape and depth you're looking for, whether you want it heated, etc. Both of you are trying to become comfortable with the other through such questioning. Obviously, you want to feel good about the contractor's competence and honesty before you let her start excavating your backyard and pouring concrete.

But the contractor is also evaluating you. She wants to evaluate whether you are a serious potential customer or are just "kicking the tires." She's also looking to gauge how realistic your expectations are and whether she will be able to satisfy your demands given her level of current and expected work. Based on this first meeting, you might decide that you'd rather have someone else build your pool or perhaps that you really don't want a pool at all. By the same token, the contractor may conclude you're not the sort of customer she wants—you might come across as someone who will be constantly complaining about something and be impossible to satisfy—or that she cannot accommodate the schedule you want because of her present workload.

An extended discussion of the work to be performed is part of the first meeting. The contractor might point out that the sort of pool you say you want might be difficult (or impossible) to install or more expensive than you anticipate. The contractor will ask you questions to get a more precise definition of what you really want (you may find yourself starting to ask for something considerably different than you did at the start of the meeting) and offer alternatives for your consideration. Instead of just a swimming pool, the contractor might suggest a combination swimming pool/whirlpool spa. Or she might suggest just installing a whirlpool spa and patio area since, in her opinion, your backyard is too small to contain a pool and its associated water treatment system.

At the end of the meeting, the contractor will offer to prepare an estimate for your consideration. During the preparation of the estimate, she may phone you to request additional information or to seek clarification on some points. Eventually, you will receive a written estimate describing the work to be performed, the schedule for its performance, and the amounts you will pay at different times during the schedule.

If the price and schedule meet with your approval, all you have to do is return a signed copy of the estimate form to the contractor along with a check for your down payment and work can begin on your pool. However, something about the bid—the price, the schedule, etc.—may not be acceptable. You go back to the contractor and voice your objections. Instead of lowering her bid, however, she starts asking you more questions and forcing you to make choices. Do you really want a whirlpool spa as part of your pool? You could save $xxxx by eliminating that. You could also reduce the bid by specifying a smaller pool or eliminating the heating system. She can also reduce her bid by taking a longer time to install the pool. Or, if you want the pool ready by summer, she can speed up the installation schedule, but she will have to put some workers on overtime and that will force her to increase the bid. After some negotiation, she prepares a revised bid and submits it for your acceptance and down payment.

When the contractor begins working on the pool, she does so largely independent of you. She supplies all tools and materials, obtains any construction permits needed, hires all necessary workers and subcontractors, determines working hours, eats lunch when she wants to, and decides (subject to the schedule contained in her bid) when work is finished and ready for your review. While you retain final approval over all work done, you do not supervise her efforts or those of other workers and subcontractors involved. In fact, she will resent any attempt on your part to manage her. And, if you're smart, you'll stay out of the way of her and her workers until it's time to review the work. After all, a large chunk of what you're paying for is her expertise in building swimming pools. Why pay for that expertise if you're not going to take advantage of it?

Finally, the big day arrives and your new pool is "presented" to you! There will be a final walk-through and inspection with the contractor, and she will probably guarantee her work for a certain period of time. You may get training in how to perform some tasks, such as draining or filling your pool. Finally, you sign a form indicating your approval of the work performed and write a last check payable to the contractor. Your kids have a new pool to enjoy and the contractor continues working on pools for other people.

This little tale has been a template of how virtually all independent contracting relationships work. No matter what skills you possess and what work you perform—building swimming pools or designing computer circuit boards—your relationships with your clients will follow a remarkably similar pattern.

Tests For Being An Independent Contractor

Whether you're really an independent contractor or not is not just a question of semantics. The Internal Revenue Service and the departments of labor in most states will classify an employee as a permanent, full-time employee instead of a contractor unless certain tests are met. If you don't meet these tests for classification as an independent contractor, you could be subject to a variety of requirements, such as mandatory withholding from payments you get from client companies. While not all-inclusive, here is a list of the main factors in determining whether you are an independent contractor as far as the law is concerned. The more of these tests you can successfully meet, the more likely you are to be legally considered an independent contractor:

✦ **You represent yourself as an independent business.** You don't have to be formally incorporated to meet this test (although that certainly would be enough), but you do have to operate as an independent business. Factors such as use of a registered business name, obtaining a local business license, membership in associations for independent businesspeople, and a listing as a business in the yellow pages of a telephone directory are all evidence of your desire to be considered as an independent business.

✦ **You actively seek business.** This is more than just registering with a job shop or temporary employment service. Instead, it means you place advertisements, prepare and mail out promotional materials about yourself and your services, have business cards, make telephone calls to potential clients, attend trade shows and events, and otherwise seek to make yourself known to potential clients.

✦ **You have a separate place of business.** An office in your home can meet this requirement, but the IRS and other agencies are taking a tougher stance on allowing home offices for tax purposes. Renting office space outside your home, complete with separate business telephone lines and a business mailing address, clearly would meet this test. This is perhaps the most common pitfall for "contractors" who are actually temporary employees, and is the factor that the IRS and most state labor agencies look at first. Meeting with clients at their place of business or doing some work at a client's place of business will not automatically cause you to fail this test, but if you have no separate business premises of your own where you do a substantial part of your work you will have trouble establishing that you are really an independent consultant.

✦ **You provide your own tools.** "Tools" means anything normally used in your business or profession, such as computers and software, arc welding equipment, pencils and paper, or a helium tank for inflating balloons for a kids' birthday party (clowns are often independent contractors too!). This is another test that trips up a lot of "contractors" who are actually temporary employees. If you perform tasks using equipment or tools furnished by the hiring company, you're on thin ice legally.

✦ **You have multiple clients.** The best way to meet this test is to perform services for more than one company simultaneously over an extended period of time. If you work only for one company for extended periods, you run the risk of being considered an employee instead of a contractor. It's better to work simultaneously for four different companies over the course of a year than it is to have consecutive three-month-long assignments with four different companies where you work exclusively with one company during each three-month period.

✦ **You have the opportunity for a profit or loss on each assignment.** Employees get paid a set rate for their labor, usually for a given unit of time such as an hour or month. Since you're a separate business when you're an independent contractor, you might actually suffer a loss on an assignment if you underestimate the time it will take you or your expenses connected with it.

✦ **Your fee and other terms and conditions of assignments are determined through direct negotiations with the client.** The other conditions include such things as the schedule for completion of the assignment, standards for acceptance, and review process. These conditions are made a part of the written agreement for a project.

✦ **Written agreements are made for each assignment.** This does not refer to a general employment agreement in which tasks and duties are specified only in general terms. Instead, this means that each task—whether designing a new computer board or installing a swimming pool—is governed by a separate written agreement describing the particulars of that task.

✦ **You determine how the task is to be performed and work without direct client supervision.** Going back to our swimming pool example, the contractor will determine which tools and methods are most appropriate for fulfilling the written agreement and how those tools and methods should be used. The contractor can hire employees for the project or subcontract tasks to others. Clients of independent contractors only have an overall approval of work performed at intermediate and final stages of the project; they do not supervise the contractor on a daily basis.

✦ **The client cannot unilaterally alter the agreement for a task.** Every employee—whether full-time, part-time, or temporary—knows what it's like to have their job assignment arbitrarily and capriciously changed. Since independent contractors work under a binding agreement, clients seeking to change the tasks to be performed must ask the contractor to renegotiate the agreement. An independent contractor can refuse any task not covered by the agreement with the client, and the client has no recourse. As a practical matter, agreements are commonly renegotiated, and a willingness to do so is an important part of any ongoing relationship with a client. However, neither the client nor the contractor can just walk away from an agreement they don't like without running a profound risk of being sued.

The swimming pool contractor we discussed in the last section easily meets all these tests for determining if someone is an independent contractor. In the remainder of this book, we're going to assume that you will meet all of these tests as well—because if you don't, you're probably not a true independent contractor!

What About Benefits?

One frequent complaint from temporary employees is that they receive no benefits such as health insurance, life insurance, retirement benefits, or paid vacation. You might think independent contractors would be in the same boat, but most do have insurance, a retirement plan (typically an individual retirement account—IRA—or a 401K plan), and take vacations—sometimes lots of vacations. Companies figure the cost of employee benefits into their budget for personnel expenses; contractors build the cost of such benefits into the fees charged clients.

Actually, independent contractors are in a better position as far as benefits are concerned than most permanent employees. You can choose the insurance plans that make the most sense for you, not the ones that are best for the company. Individual retirement plans can also be customized for your best interests and you can control how those assets are invested, unlike participants in a company-sponsored retirement plan. You're not stuck with some arbitrary, inflexible vacation or holiday policy, either. You also determine when you take vacations and for how long, as well as which holidays you'll observe. (Shouldn't your birthday be a holiday?)

The Mindset Of An Independent Contractor

Every successful independent contractor we have ever known has had certain personal and attitudinal characteristics in common with other successful contractors. We've seen some otherwise well-qualified people fail miserably as independent contractors simply because they lack those characteristics. We'll even go so far as to say that you can't succeed as an independent contractor unless you possess most of the characteristics we'll discuss here.

Let's go back to our swimming pool example. Suppose one of the contractors you talked to had a grumpy, who-gives-a-damn attitude and complained about how tough things are in the swimming pool installation business these days. Suppose another contractor was cheerful, upbeat, and mentioned how things have been a bit slow in the pool business lately but it's now looking up. When they submitted their bids, the first contractor underbid the second by about 10%. Who would you give the job to?

> ***"Not everyone is a good fit inside a corporation. I've left the corporate world twice now—the second time I took a corporate job because I was buying a house and felt that I needed a 'stable' income. That only lasted a couple of years. I don't see myself ever working for another company again, other than on some sort of outside retainer relationship. Once you've been your own boss, it's hard to answer to higher-ups who know less than you do."***
>
> **Lynn Edwards, Bookmark Book Production and Prepress Services, San Diego, California**

Probably the second one—the cheerful, upbeat one with the higher bid. You wouldn't do so because you like spending money, but because you would feel much more confident of getting good value for your money. The first contractor—the negative one with the lower bid—smells like a loser. Sure his bid is lower, but you would suspect that he wouldn't be especially motivated to do a good job for you. You would probably rationalize away the lower bid with things like "you get what you pay for" or "he's probably desperate for the work, that's why he's so cheap."

Of course, you might be completely wrong in your reasoning. The negative, cynical contractor might really be capable of doing a first-rate job at a bargain price. But he would never get a chance to do that work because you were turned off by his personality (or lack of same). The same thing applies to you as an independent contractor. Clients are not buying just your skills or services; *they're buying you.*

Using an independent contractor always involves some degree of risk for a client. This risk can be expressed in terms of money, time, opportunities, or even someone's job if you don't perform as you promise. A large part—maybe the biggest part—of a decision to use you as a contractor will be the client's emotional reaction to you. A client has to feel confident that you're going to reduce his or her problems, not add to them. All the qualifications and experience in the world won't do you any good if clients don't feel they can trust you to get the job done right.

You might think the previous paragraph was a case of stating the obvious, but it is truly amazing how many would-be independent contractors never grasp those points. They go their merry way, oblivious to the signals they're sending clients and potential clients, and then wonder why their introductory meetings with clients are so brief or their proposals and bids are never accepted. Too many new independent contractors expect to get assignments by virtue of their "obvious" skills or high positions they've held; they neglect to make potential clients feel comfortable about working with them. Studies have shown that persons who have held high executive positions in companies actually tend to have a more difficult time making it as independent contractors than do those who enter contracting from lower levels in corporations. We don't know why this is so, but we suspect one big reason is that spending too much time as an executive can cause you to forget how to make others feel comfortable with you—or why such a thing might be important!

You don't have to enroll in a charm school or undergo psychoanalysis before setting up shop as an independent contractor, but you do have to take a hard look at what kind of person you are and how others see you. We went through a similar exercise ourselves when setting up HighText. In dealing with other independent contractors as peers and as their clients, we've noticed that successful contractors tend to share some common characteristics. Try answering each of the questions in the following list. Some of the items on our list involve how others react to you, so it might be useful to ask others you've worked with (not your friends, who tend to tell you what you want to hear) about how others perceive you. If you're honest with yourself, you'll have a good sense of whether you have the personal attributes you need in independent contracting:

✦ **Do you really like what you're planning to do as an independent contractor?** Even at the professional and technical level, many people don't like what they do for a living. This explains why mid-life career changes are so common. If you don't truly enjoy what you plan to do as an independent contractor—problems and all—and if you don't look forward to doing it every day, then don't try to do it as an independent contractor. It's no secret when someone would rather be doing something else, and clients and potential clients can sniff that out in a hurry. Moreover, if you don't like what you do, you won't be able to summon the necessary enthusiasm and energy to keep going when things get rough or you have problems on assignments. The urge to avoid dealing with tasks and problems, either overtly (going to the golf course when you should be working) or in a more subtle fashion (endlessly researching a problem instead of actually dealing with it) will be overwhelming. Such behavior is common in large corporations, but it will kill you as an independent contractor.

✦ **Is your attitude toward your work and life generally positive?** Few things are more excruciating than having to be associated with someone for whom the glass is always half-empty. You know these types and how tiresome they get . . . they have problems they complain about but never try to solve . . . the only time they perk up is when some new effort fails (because they knew it would never work) or when it's time for lunch . . . they talk in a dull monotone, sit with their shoulders stooped, and say "I don't care" often. Believe it or not, we have sat through a few presentations from independent contractors who acted in such ways! Needless to say, we get such people out of our offices ASAP.

✦ **Are your skills current?** You may have years of experience in a field, but organizations are typically looking for contractors whose skills and knowledge are as current as possible. At HighText, we need contractors with skills in computer-based book production, including familiarity with popular computer hardware and word processing, illustration, and page layout software. However, we still get inquiries from contractors whose skills are still based on conventional typesetting and paste-up page layout, and we can't use those people at all. A combination of years of experience coupled with the latest skills is tough to beat. And a knowledge of current methods and technologies can compensate for a lack of experience—if the methods and techniques are really new, then no one will have extensive experience with them!

✦ **Is doing a good job important to you?** This doesn't mean you want to do a good job because it keeps you out of trouble or someone pays you a compliment. Instead, is being a person who can do a task competently and professionally an important part of how you see yourself? Do you get real satisfaction out of doing something well even if no one else notices? If you do less than your best, do you feel bad about it afterward even if everyone else is happy with the results?

✦ **Can you accept responsibility—and blame?** This is a trick question; everybody likes to say they can accept responsibility. But when something goes wrong, their actions tell a different story. Somebody else didn't do what they were supposed to do . . . something didn't arrive in time . . . no one told them . . . or whatever. It's a rare person who can really accept full responsibility for an unhappy result and admit "I screwed up; it's my fault entirely." Yet accepting full responsibility for an outcome is a requirement for an independent contractor.

✦ **Can you see things from somebody else's perspective?** As an independent contractor, you have to look at a situation as your clients do. This doesn't mean that you lose your capacity for independent thought and judgment or agree with everything a client says or does. However, it does mean that you have to be able to understand what is motivating your clients and what their needs are. You don't have to be psychic to do this, but you do have to be a good listener, observant, and willing to ask questions of clients.

✦ **Are you willing to take the initiative in dealing with problems and suggesting solutions?** Organizations usually don't hire an independent contractor to implement a solution they've devised. Instead, they want contractors who can analyze problems and devise solutions for them. Organizations don't often suggest additional tasks that a contractor could undertake; that's something that an independent contractor who's willing to seize the initiative has to do.

✦ **Are you resilient?** Independent contractors must be able to handle disappointment. No matter how good you are, you will only be able to turn a small percentage of potential clients into actual paying clients. You'll knock on a lot of doors that refuse to open, have assignments that looked certain evaporate into nothing, and find yourself forced to deal with a lot of unexpected events. If you're the sort of person who concedes defeat and gives up quickly, then independent contracting isn't for you.

✦ **Are you a self-sufficient person?** As an independent contractor, you will have to provide your own moral support, question your own assumptions, and take a detached look at your plans and performance. While networking among independent contractors is helpful and growing, more often than not you'll be on your own in dealing with situations and problems you encounter.

✦ **Are you comfortable with risk?** This is perhaps the most crucial question of all. Independent contracting is not for those who are risk-aversive. It has what investors like to call "high upside potential" but also carries with it the possibility of failure with big personal, financial, and career consequences. This question doesn't imply that foolhardiness or rash behavior are desirable in a contractor. Instead, are you able to handle the sort of risk where your efforts are the major factor in determining whether or not you succeed? Remember that risk is an inherent part of any job situation, as some people who thought they were set for life because they had a job at IBM or Kodak found out in recent years. For an independent contractor, risk is out in the open instead of hidden out of sight. Some people prefer it hidden, and they should probably remain full-time employees.

If you can answer these questions in the affirmative, that's no guarantee of success as an independent contractor. But if you *can't* answer them affirmatively, that is almost a guarantee of failure.

Independent contracting is not the right career option for those who like to have other people make significant decisions about their careers for them. But if you prefer to drive yourself instead of being driven by someone else, and have confidence in yourself and your abilities, then you can prosper on your own.

Chapter 2
The Product And The Process

Some people think you need to know a lot of people in different organizations willing to give you assignments in order to get started as an independent contractor. However, that's not true; in fact, starting out by relying on your existing contacts can retard your progress as a contractor or even eventually cause you to fail due to a lack of assignments. The reason why is that your existing contacts—no matter how numerous and influential—will never be able to provide enough steady work to make independent contracting viable for you. Eventually, you'll need to know how to develop new contacts in other organizations and, perhaps more importantly, how to convert those new contacts into paying clients. We'll call the techniques you'll use to locate organizations who need your services and get assignments from them *the process.*

But before you can sell your services to an organization, you have to decide exactly what you're trying to sell. You are *the product.*

Who Are You And Who Needs You?

Many new independent contractors try to find assignments the same way they would look for another permanent job with an organization. They mail out resumes and wait for organizations to call them.

The problem is that it doesn't work that way. Companies aren't going to try to figure out what you can do for them. Take it from us—we get scads of resumes at HighText, and we don't have the time (or interest) to figure out what those would-be independent contractors should be doing with their careers or lives. You have to figure that out yourself and then make the case to prospective clients that they need you.

You might think the way to make this case would be to stress your education, skills, experience, and special talents in your presentation to a prospective client. That doesn't hurt, but it misses the point of what you're really trying to sell.

You are selling solutions to client problems. You may have a superb education, current skills, years of experience, and be a talented musician or expert mountain climber to boot. But none of that matters unless you can make a clear connection between your personal attributes and the solutions to problems facing an organization.

This is something that is not easy for many new independent contractors. After all, stressing things like previous jobs and your education is how you land full-time positions with most organizations. Looking for a contracting assignment is different, however. You have to start thinking in terms of the *value* you offer to an organization. You can't go to a company as an independent contractor and say something like "Here is what I've done in the past and where I went to school; is there anything I can do for you?" Instead, you have to say something like "I've been successfully helping companies like yours handle the same type of problems, and I can do so easier and less expensively than alternate solutions you're considering. Give me the assignment."

A good way to figure out how you offer value to an organization is to use the same technique companies use to determine the relative value of their permanent employees and how much to pay them—a *job analysis.* This means the company breaks down each job into a set of tasks and tries to determine how much the successful performance of each task is worth to the organization. You don't have to do this for each job you've held. Instead, analyze what you're interested in doing as an independent contractor in light of the tasks you actually performed in your previous jobs. This can be a revelation, as you might discover that your previous job titles often don't reflect what you have actually done in your career, especially if you've held positions with ambiguous titles like "director," "coordinator," or "administrator." A job analysis is also useful because it helps you focus on those tasks you've performed in previous jobs that are really useful to other organizations instead of on tasks that are particular to your previous employers but of no interest to other companies.

Let's take a real-world example. While this book was being written, one of the authors had several long conversations and meetings with a friend who had spent the previous 14 years as a "buyer" for a multibillion-dollar national retail chain. His specialty had been electronics items, and he was trying to decide what he could do as an independent contractor. During these conversations, we "job analyzed" his experience and found that his previous title of "buyer" was inadequate to describe what job tasks he had performed and would be of interest to other organizations on a contract basis.

Our friend's previous job title should have been "merchandising manager," for he did far more than just buy electronics items. Here is a list of *some* of the tasks he routinely performed:

- selected which electronics items would be carried in retail stores
- negotiated prices, quantities, and delivery schedules of items from outside suppliers
- set retail prices for all items in his product area
- worked with suppliers' engineering departments on incorporating changes in the design of items

- worked closely with suppliers that custom-manufactured items for resale under the retailer's brand names

- approved all advertising materials for his product areas; gave substantial input to advertising people for preparation of such materials

- attended major trade shows such as the annual summer and winter Consumer Electronics Shows in Chicago and Las Vegas

- had final approval of all instruction manuals for products as well as approval of packaging design

- gave presentations to retail store managers for upcoming new products and helped retail store managers with problems relating to his product line (this person had once been a store manager for this retail chain)

- traveled to Japan, Korea, and Taiwan on a regular basis to meet with suppliers and negotiate contracts with them

Over the years, the items in his product category had ranged from individual components (such as fuses and integrated circuits) to such relatively complex devices as security systems, microwave oven leakage detectors, and electronic home blood pressure checkers. Suppliers had included U.S. companies as well as foreign ones.

Note that the preceding list of tasks did not include such things as preparing reports for senior management, supervising subordinates, or other functions related to "servicing the bureaucracy" at his previous employer, as it was unlikely that other organizations would place a value on having such tasks performed.

Once you've gone through and determined just what tasks you've done in the past, the next step is to ask what sort of problems those tasks solve. In the case of our friend, here are some questions that potential client organizations engaged in the manufacture and sale of electronic products might be asking themselves:

- What is the market for certain electronics products?
- What do we do about advertising for our products?
- How can we get the instruction manuals done and the packaging designed?
- We want to manufacture one of our products in Asia; how do we find a supplier, negotiate a contract, and oversee its implementation?
- How much should we charge for our product?
- What kind of support will retailers of our product need?
- What trade shows do we need to attend, and what do we do once we're there?

Since our friend demonstrated a capacity to solve the problems behind such questions, he stood a good chance of landing several contracting assignments if he could locate companies asking those questions. The "product" he can successfully sell as an independent contractor is *product development, production, and marketing services for small manufacturers of electronic devices.*

You have to develop a similar concise, easily communicated definition of what you're selling. One reason is that prospective clients will almost always ask you—usually at some point early in your first conversation with them—a question like "What is it that you do?" If you can't immediately give a short, to-the-point answer, you're in big trouble. You also need to answer that sort of question for your own benefit, because you're in even bigger trouble if *you* don't know what it is that you do!

Developing A "Product" From Your Experience

You might think the previous example was too easy, since our friend had a clear background and set of specialized problem-solving skills. But almost any job-related experience can be made into a "product" you can sell to organizations on a contract basis. Let's look at some typical job titles for permanent employees and see how the tasks performed as part of those jobs can be sold on a contract basis.

✦ Credit and collections manager

Most larger companies have at least one person responsible for reviewing credit applications from other companies, making decisions on whether to grant credit and on what terms, and following up on delinquent accounts to collect money owed. Smaller companies also sell to other companies on credit even though the application processing and collection procedure is seldom the exclusive duty of one person; typically a person handles it along with other duties. Yet smaller companies need to be even more careful in granting credit and getting paid promptly than large companies. Late payment of a $50,000 invoice is merely annoying for a multimillion dollar company, but for a small company it can mean whether a payroll is met or not.

Given this, an obvious opportunity for independent contractors would be to perform credit review and collection functions for smaller companies on a contract basis. (As we'll discuss later in this book, this is an ideal situation for some sort of retainer relationship.) Another opportunity might be to approach things from the opposite end and advise smaller companies on how to secure and maintain good credit. Another possibility is to act as an advisor to cash-strapped companies trying to reach an agreement with their creditors to reschedule payments. Companies with an in-house credit and collections department may need the services of someone outside to solve problems that cannot be resolved by their existing staff. For example, a company may find itself with excessive bad debt or slow payment problems that their in-house personnel cannot solve. An independent contractor might be the best way for that company to find out what's going wrong and how to fix it.

✦ Bank loan officer

Larger companies have financial officers whose expertise includes obtaining outside financing, such as from banks, on the best possible terms. Smaller companies need financing as well, but the world of dealing with banks and other financial institutions is completely alien to many entrepreneurs. Such people would welcome a contractor who could guide them through the banking jungle. Services you could provide might include determining the capital requirements of small companies, the type of financing best for the company (such as a line of credit versus a fixed-term, fixed-amount loan), helping companies prepare the necessary documentation to support a loan application, and assisting companies in negotiating the best terms and conditions for any loan offered. You might be able to take some of your compensation as a percentage of any loan received by the company.

✦ Teacher

There has been a steady growth of instructional programs, usually of a remedial nature, outside of the traditional educational system. Any skilled teacher will have numerous opportunities to apply those skills as an independent contractor.

For example, many corporations have been forced to teach basic mathematical and reading skills to employees in order to maintain a workforce able to handle the requirements of modern jobs. Numerous former public school teachers are now making more money and having fewer headaches as trainers for such companies. Even elementary school teachers are able to contract their skills to companies because of chronic problems experienced with adult reading and mathematical illiteracy.

If your background is the humanities and social sciences, you can easily transfer your skills to conducting training programs and seminars for managerial and professional employees on such topics as sexual harassment prevention, dealing with diversity in the workforce, effective communications skills, human relation skills, and similar "soft" business skills that companies are recognizing as important in building effective organizations. For some of these topics, you may need to do some additional research or preparation. That's okay; what you're really selling is your ability to teach the essentials of a subject effectively, not necessarily your exhaustive knowledge of the subject.

If your teaching skills are in a "hard" area like mathematics, chemistry, physics, foreign languages, computers, or applied business (accounting, etc.), then you're really in luck. Companies of all sizes have an almost constant need for employees trained and re-trained in such areas. This is especially true when companies have laid off employees; since there are fewer employees, companies need to have their remaining employees as well trained and broadly skilled as possible. You can find such opportunities teaching a number of employees from a single company or in holding seminars attended by employees of several smaller companies.

We personally know of former public school teachers who have started their own businesses exclusively to teach foreign languages to business people, including weekend "crash courses" for people who are traveling to a foreign country on short notice. In California, minor traffic violations can be removed from one's driving record by attending a "traffic school" licensed by the state. These traffic schools cover basic driving safety and techniques, and many such schools are conducted by former school teachers.

✦ Human resources professionals

Small companies usually can't afford a full-time employee handling their personnel and human resources functions, but they still must comply fully with all laws concerning their relationships with employees. Small companies also must get the most out of their investments in their employees. As an independent contractor specializing in human resources, you can help smaller companies comply with all relevant employment laws (such as eligibility for overtime pay, nondiscrimination in hiring, or terminating employees), perform job analyses and classifications to determine appropriate pay scales, establish and maintain personnel record systems, and set up training programs to help companies get the most productivity out of their employees. With larger companies, you can often be the "outside expert" brought in to help with specific problems or tasks.

✦ Retail store manager

Each year thousands of retail stores are opened across the country by people with no retail store experience. Operating a retail store seems simple enough to such people—*you just open the doors and start taking the customers' money, right?*—but soon these new retailers find themselves in desperate need of assistance in running their store. If you have had some successful experience in managing a retail store, you have skills you can sell to those running a retail store or thinking of doing so. You can advise those people in such areas as site selection, merchandise selection, how to order merchandise through distributors and wholesalers, hiring and managing employees, customer relations, security, advertising and promotions, inventory management, and numerous other issues retail storekeepers must deal with. Your knowledge of such matters could be sold on a one-time basis—for those just opening a store or who have one that isn't performing as expected—or on an ongoing basis to assist in daily operations. (Managers of successful restaurants, "bed and breakfast" inns, etc., can also sell their skills and experience to newcomers to those businesses.)

These examples show the different ways your skills and experience can be packaged as an independent contractor. The key point to keep in mind is *what you offer to do is dictated by the needs of your customers.* As we'll see later, your customers may not be consciously aware that they need your services (few people who are incompetently managing their retail store are aware that they are doing so, for example). One of your tasks as an independent contractor will be to make clients aware they have problems you can solve without offending them.

The Wrong And Right Way To Get Started

If you wait for potential clients to contact you—or rely strictly upon an existing network of contacts you developed as a full-time employee—then you will eventually fail as an independent contractor. To make it as an independent contractor, you *must* initiate contact with potential clients and continue to do so throughout your career.

There's nothing wrong with placing ads soliciting clients in trade publications, printing and distributing brochures, or hanging out a shingle in front of your office. Those are all elements of a good campaign to locate clients, but such things are insufficient by themselves to generate enough business to let you keep the doors open. It is up to you to target companies that might be able to use your services, contact them, and then convince them to use your services. The main tool you'll use to do this will likely be the telephone.

This requirement to take the initiative in contacting potential clients upsets many potential independent contractors—and not doing so is the biggest failing of most unsuccessful ones. Some people compare it to prostitution or selling used cars. To avoid having to deal with this reality, many go running to a job shop. However, the job shop simply does the same thing on their behalf and as compensation takes a healthy cut of what the company is willing to pay for a service. The job shop is doing what you can—and should—be doing for yourself. And it's not difficult at all; it just takes a telephone and some practice.

Some new independent contractors expect to get all the work they need from people they already know. It's not uncommon for many managerial and professional employees who are laid off to leave with some sort of consulting or contracting agreement, which is a *de facto* form of severance pay. It is often possible to have a nice income the first few months as an independent contractor through such arrangements and picking up assignments from people you've worked with in the past. But it won't last. Agreements eventually expire, contacts move to other companies, and sometimes there are just no tasks that a company needs performed by outside people. The income stream dries up, and the pseudo-contractor starts to get desperate, making anguished phone calls to contacts that begin "I really could use some work . . ." But the contacts can't help; they're running a business, not a charity.

This doesn't mean that you don't exploit any agreements you might have with a company or that you don't call people you know. And it doesn't mean you won't get repeat business from satisfied clients, referrals from satisfied clients, or even retainer agreements from satisfied clients once you're up and rolling as an independent contractor. What it does mean, however, is that you can't rely totally upon any of those. *The ability to locate new clients for your services is absolutely essential for your survival as an independent contractor.*

This is important throughout your career, not just when you're starting out. No matter what you do—no matter how well you perform your tasks or how well your customers are satisfied—*you will lose a certain percentage of your existing business every year.* The requirements of your clients are constantly changing, and so will their need for you. You can lose a client for reasons that have nothing to do with how well you do your work or how happy clients are with you; projects are completed or abandoned, new permanent employees are hired to perform the same tasks you do, or budget cutbacks are ordered. Experienced independent contractors don't get upset about this, for they know it is as inevitable as the sun rising in the east. As a result, most independent contractors—usually the most successful ones—devote about 20–25% of their time to the search for new clients and business.

It's common to be uncomfortable with the necessity to initiate contact with potential clients. We've been taught from childhood that we're not supposed to toot our own horn or call attention to ourselves—and pitching your services to potential clients is definitely a case of tooting your horn loudly. Some people are unable to handle rejection well, taking it as a judgment on their worth as an individual, and don't want to risk being rejected after initiating contact. (This may be another reason why high level executives often have trouble as independent contractors; it may have been a long time since they last experienced personal rejection over a business matter.) Others have a notion that if they're really good, other people will find that out without any effort on their part. Contacting potential clients is also a violation of the employer/employee model most people are comfortable with (after all, you don't go into your boss's office and ask for work; it's *his* job to tell you when there's work to be done). And some are just too shy to comfortably call up strangers and start looking for work. Regardless of the reasons, however, you have to actively seek out new clients when you're an independent contractor.

San Diego, CA

Hightext Publications Inc.

San Diego, CA

Dear Sir or Madam:

I am a technical writer and editor available for freelance assignments. For the past ten years I have worked at General Dynamics and Science Applications International Corporation (SAIC). I have experience with government documentation, software training and user materials, and proposals. In addition to my professional experience I have a degree in English. I am confident I could handle any writing or editing assignment you might have.

Please feel free to contact me during the day at , or in the evening at If you would like to see writing samples I would be happy to send them to you. Thank you for your time and consideration, and I hope to hear from you soon.

Sincerely,

Figure 2-1: A complete waste of a perfectly good stamp! This is an actual solicitation letter received at HighText. The writer of this letter obviously didn't care; why should we?

Having to seek out clients actually has several advantages. For one thing, it greatly reduces your competition. If you've ever placed a "help wanted" ad in a newspaper, you know how such ads can attract a flood of responses. If companies ran "independent contractor wanted" ads, you would be competing against dozens or hundreds of other contractors. When you are the one that makes contact with a client, you often find yourself without any competition for an assignment! By initiating the contact, you play an active role in identifying problems facing the client and developing solutions. In effect, you play a big role in creating assignments for yourself. And since you are the one that starts the ball rolling by contacting potential clients, you're often in the driver's seat in directing all aspects of the contracting relationship.

"In my line of work, the only way you can expect to make a really good living is to do it on your own. When I first left the company I worked for, I came home, set up my studio, and sat back and expected the phone to ring. I finally realized that no check was coming in, and that I needed to get off my butt and market myself. I'm a little shy and marketing is hard for me—it doesn't come naturally to me to tell everyone how great I am and beg for work. I had to overcome my 'breeding' that taught me that such behavior was rude and obnoxious. My first job actually came through a referral from my wife—she's turned out to be one of my best agents."

Brian McMurdo, graphics artist,
Valley Center, California

Initiating contact with potential clients and developing working relationships with them can be done according to some simple, proven steps. At the beginning of this chapter, we referred to these techniques as "the process." While you'll make some modifications to the process to better reflect the services you offer and the organizations that use them, it is still the basic framework used by almost all successful independent contractors.

The Process: How Independent Contracting Really Works

Looking for new clients involves more than just mailing a brochure or calling a company on the phone and asking them if they have any projects for an independent professional in your field of expertise. In fact, if you try blindly try a "mass mailing" or call companies asking if they need an independent contractor in your specialty, you will probably get nothing but polite turndowns from those companies, although several will probably indeed have problems you could solve for them. So how do successful independent contractors manage to locate potential clients that need their services and convert them into paying customers?

How to do that makes up the rest of this book. But the techniques we'll discuss are based on some fundamental principles. Don't forget the following:

✦ **You have to know what it is you're trying to sell.** Presumably you've already gone through the sort of exercises we discussed in the first half of this chapter before you try to contact any potential clients. We are contacted at HighText by too many would-be contractors whose basic pitch is "I'll do anything legal for money" or who just mail a generic resume. If you haven't yet figured out what services you can perform for a company, don't expect them to do it for you—because they won't.

✦ **You have to focus your client search efforts.** You have to contact those potential clients who have a realistic probability of needing services like yours. Trying to indiscriminately contact every company in your area or an industry will just waste your time and theirs. It also won't produce much income for you.

✦ **You have to get through to someone who can make a hiring decision.** Everyone you meet with or talk to in a company might think you're terrific, but it does you no good if you can't convince the person with the ultimate authority to authorize using you. While you can't always reach this person when you first contact a potential client, reaching that person must be your ultimate goal.

✦ **Clients care about themselves, not you.** Potential clients are not going to retain you just because you're a wonderful person with a sterling record of accomplishment or have a cheerful personality. In fact, they really don't care much about you except to the extent that you can solve problems and perform tasks for them. Does that sound cold? Well, when was the last time you took a personal interest in an auto mechanic, plumber, or dry cleaner? Your clients are no different than you are in those situations. Keep the emphasis in your efforts on what your clients need, prefer, or feel because nothing else really matters.

✦ **Make your clients talk about themselves and LISTEN.** There are several good reasons for this. Since clients don't really care about you, they're unlikely to listen to you talk at length about yourself. But people do like to talk about themselves. Moreover, the more your clients talk, the more you can find out about themselves and their problems. That lets you propose solutions that you can implement. Most clients will probably also reveal more about themselves than they realize or intend, information which will usually be to your advantage.

✦ **Ask for it instead of waiting to be asked.** If you want to meet with a client to pitch your services, ask for the meeting. If you can perform certain services for the client, draft a proposal for their consideration instead of waiting for the client to decide if there's anything you can do for them. Name your price instead of letting them figure out how much your services are worth.

✦ **Get it in writing.** You're a "contractor," and that means you put all agreements between you and a client into writing, even if it's only a letter saying "This is what I understand that you want me to do." If you don't, you will sooner or later (usually sooner) be out some money.

> ***"The times I have trouble with a client always seem to be when I don't have a written agreement. As someone once said, 'Oral agreements aren't worth the paper they're printed on!' When you're just starting out, you might hesitate to ask a client to sign an agreement, because it seems too formal. But do it anyway! I have a simple, fill-in-the-blanks form in my computer. No one has ever refused to sign it, and asking for a signed agreement makes it clear that you're a professional."***
>
> **Lynne Friedmann, public relations consultant, Solana Beach, California**

✦ **Get used to "no" and don't pay it any mind.** The easiest way for any potential client to deal with your initial approach is to say "no," even if they really do need someone like you. A "no" might mean they really don't need someone like you, or it might only mean the person taking your call didn't know what to do next. It's possible to convert some of those negative responses to positive ones. You'll hear "no" a lot as an independent contractor, so get used to it and don't take it personally.

✦ **Act like a professional.** This means you have to supply your own motivation and direction, and that you assume full responsibility for delivering the results you promised. Trying to pass blame for a missed commitment to someone else ("The printer promised me it would be ready this morning") won't cut it.

✦ **Act like you're really independent and under contract.** If you act like a full-time employee, you'll be used—and abused—like one. The contract governs your relationship with the company, and neither of you can change the terms of the contract without mutual agreement. If the company wants you to do something not specified in the contract, you're justified in asking for more money for those new tasks. And the company is equally justified in insisting that you complete a project by the date specified in the contract.

So what is this mystical "process" that's derived from these principles? The process of independent contracting can be reduced to just four steps:

1) **Identifying and contacting new clients.** You start by locating organizations that could possibly use your services. You then make contact with those organizations and identify the person who can make the decision whether or not to use your services. When you do, you contact that person and set up a meeting.

2) **Meeting with the client.** During the meeting, you determine the organization's needs and problems, and also start the process of convincing the hiring individual that you're the one who can help them solve those problems.

3) **Presenting a proposal to the organization.** You put together a plan of action to solve the organization's problems, which includes the tasks you'll perform, the schedule for completing those tasks, and how much you'll be paid. To accept your proposal, the hiring individual just signs it in the space provided.

4) **Doing the contracted work.** Once your proposal is accepted, you perform the tasks you proposed and get paid.

And that's it! However, there are plenty of details involved in performing those four steps. Let's start getting into those details.

Chapter 3
The Client Hunt

Many of us like to think that some sort of meritocracy is in effect for professional workers—do better work than your competitors and you'll make more money than they do. However, a mediocre independent contractor who's good at locating, contacting, and securing meetings with new clients will often do better financially than a more skilled individual who's not very good at getting in front of new clients. If you're good at what you do, and are good at finding new clients, then you'll do very well indeed as an independent contractor.

What's Your Area?

Before you start contacting organizations, you have to determine the geographic area where you'll look for clients. If you choose an area that's too small, you won't generate as much income as you'd like. But if you choose an area that's too big, you won't be able to properly focus your efforts in targeting likely prospects.

Sometimes the right area is obvious, especially if you live in a metropolitan area. If you're offering commonly needed services like accounting or desktop publishing, the geographic area you'll target will probably be the city in which you live. You might decide to serve only a portion of the city because of factors like traffic or proximity to your home.

Long Distance Business

"With Federal Express and fax machines, I find that working long distance is wonderfully effortless. I know that this doesn't work for everyone, however—some people need proximity. I definitely prefer to work independently. Those clients who want to look over my shoulder every minute are not the ones I want to work for anyway."

Sara Patton, book editor and desktop publisher, Wailuku, Hawaii

Things get stickier if your skills are more specialized or if you live outside a metropolitan area. If there are few companies in your immediate area that might possibly need your services, then you'll have no choice but to expand the area where you'll try to locate clients. Suppose that your service is offering Japanese-English translation services with an emphasis on business documents. If you live in Los Angeles, there are literally thousands (or tens of thousands) of potential clients you can contact. But if you live in rural Montana, you might not have a single potential client within easy driving distance of your home. You might be forced to look for clients statewide, regionally, or even nationally. While this is not as easy as looking for clients locally, it can be done successfully. It's true companies often prefer to work with someone local, but they prefer more to work with someone who's good.

As a general rule, it's best to keep the geographic area you draw clients from as small as possible. One key reason is how much contact you'll need with your clients. A lengthy trip for an initial meeting with a client is fine if most of your additional contact can be made by telephone, fax, or Federal Express. But if regular meetings or trips to the company's facilities are required, lengthy travel can quickly eat into your profits and time available for other clients. Your travel time and expense is a cost that you will have to pass along—either directly or indirectly—to your clients. Adding in the cost of a couple hours of travel several times a week can make your price noncompetitive.

Lengthy travel also cuts down on the amount of time you actually have available for work, reducing the number of projects you can complete in a period and often reducing your potential income. And if you're tied up in travel for a couple of hours each day, this reduces your ability to complete rush projects or handle emergencies for your clients.

You can adjust the size of your "service area" as events dictate. When we began offering technical manual preparation services, we decided to search for new clients exclusively in San Diego county and in neighboring Orange county. We decided on this area because some projects would require extensive in-person contact with clients, and most locations in San Diego and Orange counties would be within an hour's commuting distance of our offices. Because of travel distance, we ruled out soliciting new clients in other areas of southern California or neighboring states despite the presence of numerous good prospects in them. However, we did have one major client in Arizona because of some previous contacts there. As HighText grew, we began

to get referrals from our existing clients to companies located as far away as Massachusetts. We based our decision whether to handle such distant clients on how much travel and contact would be necessary. We discovered that we were actually able to serve some clients entirely through nonpersonal contact using the telephone, mail, fax, or Federal Express. While we welcome such unsolicited business from "low maintenance" clients, we still continue to solicit the bulk of new clients from San Diego and Orange counties. As your contracting activities grow, you will likely find yourself in a similar situation that requires you to re-evaluate your service area.

Making A List And Checking It Twice

Direct mail specialists know the importance of a good list of prospects for their merchandise or services. They will pay big bucks to rent a list of good prospects (such as people who have ordered similar merchandise within the past year) for a mailing because they know a good product or attractive direct mail package is wasted unless you reach people who are likely to buy. The same thing is true when you're an independent contractor.

Unlike direct mail specialists, however, you'll have to compile your own list of prospects. This can be a lot of work, but it's a powerful competitive tool for you if done well. If you are offering a generic service that many different organizations can use (such as accounting or graphic arts), you might be able to get by compiling a list of prospects from telephone directories. Most contractors, however, use several different sources to compile prospect lists.

A great place to start putting together a prospect list is at your local public library, which has several good references you can consult. The following list isn't exhaustive, but it indicates some of the materials you may find useful:

✦ **Telephone directories.** Many companies publish specialized directories, such as the "business to business" directories listing only the companies in an area. While you usually won't find companies in such directories that aren't listed in ordinary telephone directories, the organization and focus of these directories can save you a lot of research time.

✦ **Industrial and corporate directories.** Several companies publish books giving information on firms by business or geographical region, such as those published by Dun & Bradstreet, Standard & Poors, Thomas Registers, etc. These publications will give you such information as main products or business of the company, annual sales, number of employees, names of key managers, addresses and telephone numbers, etc. There are often specialized directories for certain types of businesses; for example, the standard directory for the book publishing industry is *Literary Market Place.* These directories are useful for locating local and national prospects. Even more specialized directories may be available for your immediate area. For example, when HighText was launched we found directories of the electronics and computer industries in southern California invaluable in compiling our initial lists of prospective clients. You can find such directories in your local library or bookstore. Ask the reference librarian—they love to help!

✦ **Government business records.** You may have to look for these at city hall or county records buildings, although copies are often available at many libraries. Records of business licenses and business name filings are especially useful, since these will give the addresses of firms, names of the owners, and the principal business of the firm. If your services are aimed at people in your area who are starting up new businesses, government records will probably be your most valuable source of prospective clients.

✦ **Annual reports.** If a corporation's stock is publicly traded, it will normally issue an annual report. Annual reports give a great deal of detail about a company, although (since it is a report to the company's owners, after all) the best possible spin is put on the information.

✦ **Trade and industry publications**. Magazines and newspapers that focus on a specific industry are great sources of information on new companies in that industry, changes in existing companies, and trends in the industry. Such publications are a great way to locate organizations outside of your immediate area.

✦ **Local business publications.** These include local/regional newspapers and magazines as well as the business section of area newspapers. Especially valuable are business "newspapers of record," which publish extensive legal notices required of businesses such as notices of incorporation, business name filings, or business license and permit applications. These can point you to new companies that might be especially receptive to the services of an independent contractor.

✦ **Membership directories of business organizations and associations.** Almost every type of business is served by some sort of association or organization, and those organizations often publish membership directories or lists.

✦ **Trade show and exhibition programs.** Major trade shows for a given industry or business attract hundreds or even thousands of exhibiting companies. Programs from trade shows can give you valuable clues about the most active companies in a field.

✦ **Networking with noncompeting independent contractors.** If a company needs an artist, they may also need a writer. If a new business needs help setting up their accounting system, they may also need help setting up personnel guidelines and policies. Staying in touch with others who are not your direct competitors can result in such opportunities.

✦ **Attending trade shows and exhibits.** If your services can be offered to companies nationally, then attending major trade shows and other events can be worthwhile. In most cases, it's not cost-effective for you to actually buy some exhibit space and solicit potential clients from a booth. However, you can often locate dozens of companies that might need your services and the names of people to contact at those companies during a single trade show.

How big should your prospect list be? There's only one rule—it can't be too big! You can always use more prospective clients, and you never stop searching for them. Successful independent contractors devote a part of each working week to locating names of new potential clients and contacting them.

When you find a prospective client, you need to estimate how "good" that prospect is. For example, a prospect that could use your services on a recurring or continuing basis is better than a client that needs you just once. You also have to consider how likely it is that a company might need your services. If you develop software for personal computers, it's possible that an accounting or financial services firm might give you an assignment to develop custom software for their use. However, it's far more likely that you'll find assignments at engineering or technical companies where some type of software development is done as part of their normal activities. Based on your knowledge of your field of expertise and the business of the potential client, you can usually make a good guess as to whether the prospect will need you or not. However, be prepared for surprises. One of our very best clients—one that has provided us with a lot of assign-

ments at high rates—is a client that we had thought would be only marginally likely to produce any work.

The size of a prospect doesn't have much relation to the probability of needing your services. Large organizations may have multiple needs and projects underway, but may also have an in-house staff that handles most of them. Smaller firms might rely more heavily on contractors to handle their smaller number of projects.

Estimating how good a prospect is helps you make the best use of the time you allocate to contacting potential new clients. You won't be able to locate an equal number of good prospects each week, so it makes sense to contact your better prospects first and "save" your less likely prospects for weeks when you have fewer good prospects. However, when you first start out as an independent contractor, contact your *least* likely prospects first. Why? Because there's a learning curve for mastering the techniques of contacting prospective clients and setting up meetings with them. You're going to make some mistakes, and you'll probably be nervous and self-conscious at first. You don't want to blow a chance with a really good prospect because of your inexperience, so it's best to practice with prospects that aren't too likely to yield assignments no matter how skillfully you contact them.

Since your list of prospects is so valuable, you need to carefully organize and maintain it. You can either go the high-tech route—a database program and a personal computer—or the low-tech—some 3x5 file cards in a filing box. But you do need to develop a system that will let you quickly access information about your prospects and update that information. You need to record the dates when contact was made with the prospects, results of those contacts, dates when any follow-up contact should be made, names of key people, and miscellaneous comments about the contact. When you convert a prospect into an actual paying client, a separate record-keeping system is needed. Most contractors use a filing cabinet to contain copies of accepted proposals, correspondence, notes, and other written materials concerning a client or project.

The Rules Of Three

Before you start contacting potential clients, you need to know the two "rules of three."

Years ago, one of us knew a senior executive in a large publishing company who applied what he called the "rule of three" to any phone call from someone he did not know outside the company. His rule of three was simple:

Never return a phone call from a stranger until he or she calls three times.

According to this executive, only somebody who has a compelling reason to want to talk to you will call a third time after having the first two calls ignored. If this person gives up after one or two calls, this executive reasoned, the subject of the phone call wasn't that important to the caller. And if the subject wasn't important enough to place a third call, then it wasn't important enough for the executive to take time to return the call. But, as he put it, "If somebody calls me three times, then they must be really serious about talking to me. And if they're that serious, it's more likely to be worth my time to return their call."

Cold-hearted? Calculating? Not very polite? All true . . . and also a very effective way for that busy executive to make good use of his time. And this executive had a valid point: if something wasn't important enough for the caller to try three times to reach him, why should he be concerned about it?

As the years went by, we discovered another "rule of three":

Never give up until somebody tells you "no" three times.

We stumbled upon this rule of three by accident. Our early efforts at contacting prospects were a bit disorganized, and we wound up contacting some of the same clients who had turned us down a few days earlier. Much to our surprise, some of those prospects agreed to meet with us and we got assignments from some of them.

As we got more experienced in independent contracting, we learned that the first or second "no" often does not mean "no." It may mean the person you've reached is too busy to make a decision at the moment . . . or they need more time to think things over . . . or they need to get authorization from their boss . . . or they have a fear of making a firm decision. In such cases, "no" is just a delaying tactic. However, if they can tell you "no" three times, then they're probably serious about that answer and it's time to move on to another prospect.

As you read the rest of this chapter (and this book), keep these "rules of three" in mind. Don't give up trying to find out something or reach someone until you've made three tries. Give a prospective client three chances to turn down—or accept!—any proposals you make. And if you hear "no" three times, that's not always the end. People and circumstances change at companies, and you can sometimes go back to a prospect a few months later and find that your calls are taken and your proposals accepted.

Ah . . . Umm . . . Phone Manners, Ya Know?

Ever notice how so many telephone solicitors sound like complete morons? Remember how irritating it is to listen to someone mumble unintelligibly, say "you know" after every sentence, or talk in a robot-like drone? You don't sound like that on the phone, do you? Do you??

Before you make any calls to prospective clients, take a cold, hard look at how you come across on the telephone. Some friends who are willing to be completely honest with you are invaluable here! Another approach is to tape-record your side on a few telephone conversations. It can be a shock to realize how you sound to others. You probably don't need to take speech lessons, but many people could improve their telephone presence just by eliminating a couple of irritating verbal mannerisms or speaking more slowly—or more quickly if they have a pronounced drawl!

A lot of people who make cold calls to prospects come across as reluctant, defensive, and hesitant, almost as if they are ashamed of making the call. Clients will usually interpret this behavior as a lack of self-confidence, and assume it is rooted in a lack of professional competence. While you shouldn't sound like a stereotypical used car salesman, you shouldn't sound like you're embarrassed to be alive either. Speak with confidence and directness; a firm "radio announcer" tone works well. Again, this is where some objective friends can be a big help.

Avoid prepared scripts. You've probably received enough telephone solicitations to know how stilted and artificial people sound reading from a script. Unctuousness and insincerity ("I hope you're having a nice day. My name is . . .") are quickly picked up by potential clients and won't earn you many points with them. Your best approach is to be professional but flexible. Always assume the time of the person you've reached is valuable, so get right to the point of your call without a lot of extraneous blather. Don't call somebody by the first name during the first call unless they only give their first name or tell you to (showing respect for a potential client never hurts). If the person you reach is in the mood to talk to you at length, then be flexible enough to spend more time with them. And LISTEN to the person you've reached; be alert to any cues they give and be prepared to shift gears quickly based on what you're hearing from them.

No one will decide to use your services just because you come across well on the telephone. But a lot of people will decide—often in a matter of seconds—not to waste any more time with you if you sound like an oaf on the telephone.

Something is done so frequently in phone calls that we have to point out here how inappropriate and self-defeating it is. It's explaining why you so badly need some work. "I'm getting married and I need to put some money away," "Money's tight now and I really could use the work," or "I've got a kid in college" may all be true, but so what? Those are your problems. A client will give you an assignment only because they need something done by you, not because they're feeling charitable. Moreover, such phrases just tell the client you're in a poor negotiating position. Clients will hire you because you seem like the best choice to solve a problem for them, not because you sound desperate. You're selling your services, not begging for alms. Always remember that.

Finding THE Name

As we noted in the last chapter, your initial contact with a potential client has just one aim: getting through to the person who can make the decision on whether or not to hire you for an assignment. Your odds of getting through to that person and securing assignments greatly increase if you know the name of the person you want to reach.

How often do you pay attention to mail you receive that's addressed to "occupant"? The same thing happens in a company when something is addressed to "accounting manager" or "director of corporate training." Many companies won't even have someone with that job title, and even if they do they are less likely to pay attention to a letter that doesn't have a real name on it. The same thing applies—maybe more so—if the initial contact is by telephone. When you ask for somebody by job title or function instead of name, that's a dead giveaway that some sort of sales call is being made and your call may be screened out by the receptionist or secretary. If you know the name of the person you need to talk to, your odds of having them directly take your call go up exponentially.

Some corporate directories and other publications will give the name of the person you want to talk to. (Call the company anyway to verify the person is still there, though.) More often, however, you'll have to dig up the name yourself. That can be a challenge, since an increasing number of companies are reluctant to give out the names and titles of their employees—one big reason is to prevent "headhunting" firms from locating prospects they can place with other companies! This means you'll need to be a little bit creative to locate the right person to talk to.

Unless you're dealing with a really small company, the person you ultimately want to speak to is unlikely to be the one answering the phone. Instead, you'll probably get a receptionist or secretary when you dial the main number of a company. That person may be reluctant to give the name of the person you want or to put your call through if you do have a name. That can be frustrating, but realize they're just doing their job. What you must do is make it possible for the receptionist or secretary to help you find the name you want.

You start making it possible by treating the receptionist or secretary with respect and professionalism, no matter how unfriendly or inept they might seem on the phone. Being a receptionist or secretary is not an easy or high-paying job, it ranks low on the corporate totem pole, and they seldom get any positive feedback. If you treat such people with courtesy and as an equal, you will often be pleasantly surprised at how they will go an extra mile to help you. You can never tell what kind of influence the receptionist or secretary might have in a company; a remark to his or her boss about what a jerk you were on the phone could kill any chance you might have of getting a meeting.

Perhaps most importantly, you can't mislead or lie to the receptionist or secretary. Don't be dishonest if you're asked a direct question. Many receptionists have developed a pretty good sixth sense about what callers want, and you won't be able to fool them. If you give a dishonest answer, and they sense that, then you're going to have a very difficult time getting through to the person you seek. (And if you do get through, they will probably be aware that you were less than honest with the receptionist or secretary.) Make an ally of the receptionist and you'll have a valuable aide to help you navigate the corporate maze.

Okay, so you're on your best behavior and you want to start calling prospects but don't know the name of the person you should contact. If you call a prospect and ask "Could I please have the name of the person responsible for software development?" (or accounting, technical writing, product development, etc.), you might get lucky and get that person's name. More commonly, however, you'll be asked "what is this in reference to?" or be told that there is no one person responsible for that activity. To prevent this, you need to tailor your opening remarks so that the person who answers the phone will be inclined to give you the name you're looking for.

Get to Know the Secretary!

"I always make it a point to get to know the secretary or receptionist at a company that I'm pitching. They are no less important than the company president and often have influence on whether or not you get on the calendar or your calls get through. Never assume that the decision-maker is only one person. Particularly in small companies, you'd be surprised how much influence a good word from a secretary can have on landing new business."

Lynne Friedmann,
public relations consultant,
Solana Beach, California

No single set of techniques can guarantee that you'll get through to the person who can make a decision on whether or not to give you an assignment. What will work best for you will depend on the industry you're in, the services you're offering, and your personality. But here are some ideas that worked for us when we first started contacting potential clients about our technical manual preparation services:

✦ **Own up to what you're after.** Not only is this the honest thing to do, it's usually the most effective. Simply say "I offer production planning (or whatever) services on a contract basis to companies like yours, and I'd like to send the appropriate person at your company some information about my services. Who should I contact?"

✦ **Talk to a name you know in a related function.** This is a great way to bypass the receptionist. Receptionists may be trained to screen out certain types of callers, but other people in the company normally aren't. If you reach someone else in the company and they happen to know who you should talk to, they will usually be happy to tell you that person's name and even forward your call. In our case, we found that the listings for most companies in electronics and computer industry directories did not include the name of someone responsible for technical manuals and documents. However, most did give the names of managers responsible for engineering, software development, or marketing. Contacting those people really paid off for us. Often technical manual preparation was one of their responsibilities; if not, they invariably knew who was responsible.

✦ **Act like you're a customer.** This can work well if your service is something that might wind up in the hands of the company's customers. Suppose you're a commercial artist. Telling the receptionist "I have a question about the artwork on the box your product came in; can you tell me who I should talk to?" might get you the name of the person you need to speak to about getting some art assignments. Note carefully that we're talking about *acting* like a customer; don't lie if you're asked whether you are really a customer!

✦ **Call customer service.** This is a variation on the previous idea. If a company has a customer service department, they've staffed it with people used to handling all sorts of unusual questions. If they don't know who is responsible for a certain function, they often are willing to find out and call you back.

✦ **Say you're going to write a letter.** A question like "I'm writing a letter of inquiry to your company about its accounting systems; can you tell me to whose attention I should send it?" might result in the name of the person responsible for accounting in the company.

✦ **Go to the top.** The president of a company—and his or her secretary—should know who is responsible for each function in a company. If the receptionist doesn't know (or doesn't want to tell), a later call to the president's office might result in getting through to the right person. The name of the president is one name you can always get and if all else fails you can contact the president about your services. This approach is something you should do only as a last resort unless the company is small (say fewer than 50 employees). In smaller companies, the president will often have to approve the use of any outside contractors, so you might as well start with him or her.

✦ **Call before or after hours.** If you call before 8:30 am or after 5:30 pm, you may reach a hardworking employee (if it's a small company, it's likely to be the president!) who is not trained to screen calls and who is likely to know the name you're after. If they're in a talkative mood, you may also be able to get more information about the company's need for your services.

Companies in the electronics and computing industries tend to be jealous of their employees since good technical talent is always in demand by their competitors. However, we've had good success with the first option of being honest about what we're really after. Calling another person in the same company also produced good results and so did calling the president directly with smaller companies. After you've been an independent contractor for a while, you'll get a "feel" for which technique (or combination of techniques) works best for you.

Making Contact With Prospects

Now that you know who at the prospect company can authorize the use of the services you perform, you have to contact that person and arrange a meeting with them. Keep in mind that you're not likely to get any assignments from a prospect just on the basis of a telephone call or a mailing to them. Before you can make a proposal to a client and have it accepted, you have to have a meeting with the client. *Your goal at this point is just to get the client to meet with you.* The definition of success at this stage is an appointment to meet with the person you contact, not a project assignment. You're trying to get the prospect so intrigued that he or she wants to meet with you and learn more about you, so save your heavy selling pitch for that meeting.

There are two main ways to go about setting up a meeting with prospects: you can call them "cold" and ask for a meeting or you can mail them some information about yourself and the services you offer. Both of these methods have their advantages and disadvantages. Here's a quick summary:

Advantages of telephone contact

- You get immediate feedback from the prospect and can adjust your approach accordingly.
- It's easy and cheap to get started; just pick up the phone and start dialing!
- It's a quick way to contact hot prospects.

- If a prospect genuinely needs your services, a telephone call is a more effective way to get their attention than a mailing.

- Even if the prospect doesn't need your services, he or she might have useful suggestions or leads to help you locate other prospects who do.

- If they don't want your services, you at least have a chance to find out why and maybe learn something that will help you with the next prospect.

Disadvantages of telephone contact

- Rejection is common, swift, and sometimes rude; you need a thick skin to handle it!

- It's often time-consuming if the person you want doesn't answer his or her phone or likes to talk at length.

- The person you reach may need or want to discuss using your services with others in the company.

- You may have only a minute or two to grab the prospect's interest.

- Businesspeople get a lot of cold calls soliciting various kinds of business, and as a result they may have the receptionist or secretary screen out all calls from unknown persons outside the company. Worse yet, prospects interrupted in the middle of something important by an unexpected telephone call from a stranger may resent it!

- It's single-shot; the prospect may not currently need your services but may in the future. Unfortunately, they're unlikely to remember your call or name in the future.

Advantages of mail contact

- You can reach a lot of prospects—even thousands—quickly through a single large mailing.
- If a prospect doesn't need your services immediately, they can keep information about you on file.
- You control the presentation totally; you don't get interrupted by questions from the prospect.
- Written material can be better organized and more focused than conversation.

Dear New Business Owner:

Congratulations on your new business venture. As a new business, you will benefit from the assistance of a certified public accountant. Step one of running a business is setting up an accounting system to record your day to day transactions. In addition there are federal and state reporting requirements. Therefore, I am taking this opportunity to introduce myself to you and to briefly mention the professional services which I provide.

I have been engaged in public accounting for the past five years, and as a sole practitioner I am able to work closely with each individual client.

I perform year-round accounting services for small and medium-sized businesses. Depending on a clients needs I prepare audited, reviewed or compiled financial statements. This service includes bank reconciliation, quarterly returns, W-2's, 1099's, sales tax returns, and cross referenced ledgers and journals.

I prepare individual, partnership and corporate tax returns. Tax planning services are provided to minimize your tax liability.

My office is located near the intersection of Clairemont Mesa Boulevard and Clairemont Drive.

If I may be of assistance to you, please do not hesitate to call.

Very truly yours,

James R. Penaligon

James R. Penaligon

Figure 3-1: Jim Penaligon is a CPA who looks for new clients by examining records of new business licenses issued in San Diego county. His letter to us was simple but effective—Mr. Penaligon now handles all accounting for HighText!

- You aren't forced to deal with rejection personally.
- You can include samples of your work—this is especially important if your services (artwork, for example) must be seen by the prospect.
- Written material can be shared with others in the company.
- It's less intrusive on the prospect than a telephone call.

Disadvantages of mail contact

- People throw out much of the sales material they get in the mail; businesses are drowning in a sea of junk mail these days (regardless of whether it's sent first or third class, all mail companies don't expect/want is "junk" mail).
- Even if they don't throw it out, people don't read a lot of the sales material they receive.
- It's expensive.
- You can't tailor your message according to how the prospect reacts.
- You can't answer questions from the prospect.
- You still encounter rejection, even if it's not delivered personally, but you never learn the "why" behind the rejection. (Was it because they have an in-house staff for that function or was your material lost in the mail?)

Besides telephone and mail contact, there are other ways to reach prospective clients. These are not as widely used as the telephone or mail, but some contractors do use them. Here are capsule summaries and reviews of each:

✦ **"Cold" personal calls.** This means showing up unannounced at a firm and asking to speak to the person who can make the hiring decision. Good luck! This is a spectacularly ineffective and counterproductive method. The people you are trying to reach are generally too busy to drop everything they're doing and spend some time with you; more likely, they will be offended by your presumptuousness and lack of respect for the value of their time. Unless you're really lucky or in a unique situation (maybe you have some really lonely people as potential clients), this technique will make you a lot more enemies than paying customers.

✦ **Trade shows and exhibits.** If the business or industry you're offering your services to has trade shows and exhibits, these can be a good, quick way to scout for potential clients. Attendees and exhibitors are there to talk business, so your efforts to find new clients won't seem unusual or out of line. You can find out a lot about potential clients in a short time. The hiring decision maker may be attending the show; if not, someone from his or her company will be able to directly refer you to them. This can be of great help when you contact the decision makers by phone or letter; they are far more likely to hear you out if you start off by mentioning someone else at the company suggested that you contact them. The big disadvantage of trade shows and exhibits is that they are held infrequently, and you can seldom generate enough client leads to sustain you just from this route. However, using this route can be a valuable adjunct to your normal client solicitation methods.

✦ **Seminars and conferences.** These have the same advantages and disadvantages as trade shows. Since seminars and conferences are smaller than trade shows and exhibits, you have a better chance to spend time one-on-one with potential clients.

✦ **Advertisements in magazines, directories, and other publications read by potential clients.** It doesn't hurt to advertise your services, but advertising seldom generates enough business by itself to sustain you. HighText has advertised its services and made some new clients as a result, but those have been a very small percentage of our total client list.

✦ **Articles in magazines and journals read by potential clients.** These are great for your credibility; a lot of people assume that you MUST be an expert if something you wrote got into print! Sending along an article or two you've written in a mailing to a prospect or displaying a book you've written during an initial meeting can go a long way toward convincing the prospective client that you're the right person for the assignment. Unfortunately, few people will seek out you out just on the basis of an article. (If anything, people will seek you out for free advice or to sell you something.)

So there's no getting around it—you're going to have to actively solicit new clients by phone or mail in order to have a viable independent contracting career. But which is best: contact by telephone or by mail?

The answer is *both*. We've found the most productive way to land new clients is by using telephone and mail techniques in one integrated approach. This gives most of the advantages of both methods and minimizes their disadvantages.

If you had to choose just one approach, telephoning prospects would probably be best. A phone call makes a bigger impact than any mailing, and the information you get from a prospect even during an unsuccessful call (that is, one that doesn't produce a meeting) will probably be of some value to you in the future. A few years ago most successful independent contractors relied almost exclusively on telephone solicitations for new clients. However, businesspeople today are suffering from severe shortages of time. Devoting a half-hour of a business day to a meeting is a big commitment for many, and they're not as likely to schedule such a meeting with a total stranger—a meeting that might wind up a complete waste of their time—strictly on the basis of a telephone call. Even if they are clearly aware that they need someone with your skills to handle upcoming projects, they will likely want some additional evidence that meeting with you is worthwhile.

At HighText, we never set up meetings with unknown independent contractors on the basis of a phone call. We want to have some evidence that the person contacting us has at least some minimal level of skills and professionalism. While we've never decided to use an independent contractor solely on the basis of written material we've received, we have decided *not* to meet with them because of amateurish written material. Moreover, good written material makes the initial meeting more productive for you and the prospect because you don't have to devote time explaining your background and experience—the written material contains such information.

Here's how this integrated telephone/mail technique works:

1) Telephone the prospect. Briefly ask if his or her company uses the sort of services you want to sell. If the answer is yes, say you will be sending some information via the mail, and you will call again after the material arrives.
2) Mail the written material to the prospect, adding a cover letter making reference to your conversation.
3) Telephone the prospect again a week after mailing the written material and ask for a meeting.

Taking this approach has several advantages over using only the telephone or the mail. It lets you verify that the person you call is indeed the person you want to talk to (the decision-making authority may actually rest with someone else) and alerts the person to expect something in the mail from you. Of course, most of the people you call won't eagerly await the receipt of your letter. However, if there is some need for your services, if you call first your letter might get more attention than it otherwise would. Calling first also lets you tailor your cover letter in light of your conversation with the prospect.

While the initial call should be brief (unless the prospect is in a mood to talk), you can glean some vital clues about the company from a couple of minutes of conversation and incorporate what you learn into the written material. The written material you submit will probably be much better organized and coherent than you could ever be in a telephone call, and does a more effective job of selling your skills to the prospect. Unlike a telephone call, your written material can be reviewed at the prospect's leisure or passed around to other people in the company for their opinions. If there is an interest in your services at the client company, setting up an appointment for a meeting during the second telephone call should be a snap.

While the integrated telephone/mail technique has worked for us—and has worked for those independent contractors who sell their services to High-Text—it may not be best for you. Depending on the services you're offering and your prospect companies, it might be best to go a 100% telephone or 100% mail contact route. However, we strongly suggest that you start out with the integrated telephone/mail technique unless you're certain that other approaches are traditionally used for your services or with companies like your prospects. As you get more experience with contacting clients, you can "customize" your technique to reflect your personality, your services, and the clients you want to serve.

Guerrilla Flyers

"I'm fascinated with the 'guerrilla marketing' concept. First, I'll call up a company and find the name of the person who's buying art. Then I send out my standard flyer and a cover letter. I don't spend a lot on my flyers—they're laser-printed and one color—but I try to make them eye-catching, even a little bit weird. About a week or so after the mailing, I call the person and ask them out for lunch or coffee. This buys me their undivided attention for an hour or so. It's cheap advertising! I get a lot of work this way."

Brian McMurdo, graphics artist, Valley Center, California

We can't overstress the importance of having to initiate contact with prospective clients, because if you don't contact prospects you're not going to get assignments. As obvious as that is, many new independent contractors put off contacting

Figure 3-2: This solicitation from Brian McMurdo of Ventana Studio is very different from Jim Penaligon's letter in Figure 3-1, but a very different service—creative art and graphics—is being offered. Mr. McMurdo has since done numerous art projects for HighText, including the cover design and cartoons for this book.

potential new clients, usually because they don't want to hear "no." Even experienced, successful contractors hear "no" a lot, and it's something you have to learn to deal with. You don't do yourself any favors when you think about contacting prospects . . . plan to contact prospects . . . or say you'll start contacting prospects next week. If you don't contact prospects, you won't survive as an independent contractor—this is a fundamental fact of life, and there's no escaping it. When you're starting out, it helps to set fixed goals for yourself, such as contacting at least 20 prospects per day. And the process continues throughout your career as an independent contractor; successful ones devote 20% to 25% of their time to looking for new clients and gaining exposure for themselves.

Give Them Something For Nothing

Regardless of how you initiate contact with prospects, you're still faced with the problem of how to get them interested enough in you and your services to schedule a meeting. Put yourself in the shoes of a prospective client. Suppose you got a call that went something like this:

> ***"Hello. My name is Jack Roberts. I offer financial management services to firms like yours. I'd like to send you some literature about my services and hopefully meet with you in a couple of weeks."***

This isn't likely to excite too many potential clients. After all, what's in it for a prospect to get excited about? But let's suppose the call was more like this:

> ***"Hello. My name is Jennifer Roberts. I'm a specialist in financial management and work with firms much like yours. As a way of introducing myself, I'd like to send you some literature showing how your company can speed up collection of its receivables, cut bank charges for its checking accounts, and qualify for small business development grants offered by the state. After you've had a chance to review the material I will send, I'd like to meet with you and see if there are additional ways your company can make its cash work harder. Are you the correct person to receive this information?"***

Now the prospect has a reason to be interested and pay attention to Jennifer's letter when it arrives. There's a *quid pro quo* that she's establishing—she wants a meeting, but she's offering something valuable to the company in return. Even if the company decides not to use Jennifer's services, they will get some benefit from meeting with her. As a result, they are more likely to meet with Jennifer. And Jennifer knows she will have a better chance of converting the company from a prospect to a client if she can get a face-to-face meeting with the person who can make the decision on whether to use her services. That's why Jennifer is willing to give away a "sample" of her expertise to prospects in exchange for a meeting.

This is contrary to advice often given in "how to be a consultant" books. In such books, would-be consultants (or other freelance workers) are often told to avoid giving any free advice to clients or prospective clients. But this doesn't take advantage of simple human nature; people are far more willing to meet with you or pay attention to things they get in the mail if there is something in it for them. Successful salespeople in all businesses are well aware of this fact, and it's more than a little surprising that some "consultants" have been so slow to catch on to it. After all, the information you can give a potential client in your initial meeting is only a minute percentage of what you know. If the prospect finds the small free "taste" of your expertise delicious, you won't have any trouble selling more!

There's another good reason for giving prospects something that's really useful. If you send prospects material with information they value, they are more likely to keep that material on hand and refer to it. The more it is referred to and passed around the company, the more likely it is that someone there will want to use your services. When we were providing book production services for other publishers, HighText offered a small free booklet with information on producing books from disk files. We found that potential customers kept it as a reference. Even those who didn't have an immediate need for our services were constantly reminded of HighText, and many of them called at a later time.

The Written Stuff

This is where a lot of would-be independent contractors fall flat on their faces. The written material you send to a prospective client is an advertisement for yourself. Think about the advertisements you've seen in the past . . . don't you always feel that a lousy ad must mean the company (and its product) is also lousy? Or that if someone can't spell or use the English language correctly then they must lack skills in other areas too? The same thing is true when you're an independent contractor.

There are three things you must always send to the prospect: a cover letter, the material itself, and your business card. The cover letter incorporates what you learned in your telephone conversation (however brief) with the recipient of your letter. The material itself is a description of your services intended to convince the prospect that he or she needs to meet with you. And the business card lets the prospect keep your telephone number handy. Business cards are also a mark of your legitimacy; a surprising number of people won't consider you a "real professional" unless they have a business card from you!

Let's follow up on Jennifer Jones and her call about her financial management services. Suppose that the prospect had expressed some doubts about needing her services—Jennifer sensed that he felt bringing her in to advise him would be a sign to senior management that he was not up to the job—but he did want to receive the information she mentioned anyway. Given this, Jennifer might write the following cover letter:

> ***It was good talking to you today. Enclosed is the material I told you about. You may find the data on the new state business development grant program of special interest, as that program has been targeted at firms like yours. Under the program, the state offers grants up to $50,000 toward new product development in the fields indicated so long as all research, development, and manufacturing activities are carried out in this state. I've helped firms like yours apply for (and get) these grants. I would be pleased to examine your company's situation to see if it qualifies and prepare the necessary documentation to apply.***
>
> ***The services I offer companies are not a substitute for strong daily financial management, such as you provide for your com-***

pany. Instead, I show companies how they can take advantage of new or little-known opportunities to improve their cash flow and general financial health. Making small changes in a company's financial management practices can quickly add up to substantial improvements in a company's bottom line, as my clients can attest.

I'd like to meet with you so I could learn more about your company and see if there is a good fit between your company's goals and the specialized financial management techniques I'm referring to. After our meeting, I'll prepare a report for your review in which I'll outline any areas of possible improvement and discuss the techniques to make those improvements. There's no obligation whatsoever on your part, of course.

I'll be calling you in about a week to answer any questions you might have and to set up an appointment with you. I did appreciate the chance to talk to you today and I look forward to meeting you in the near future.

Jennifer tried to make her cover letter nonthreatening; she wanted to pose as little "competition" as possible to the recipient. But she also needed to suggest that there might be clear, measurable improvements as a result of meeting with her. The "report" she mentioned would actually be her proposal to the company to perform services for it, yet it also held out the promise of further useful information. She also said there would be no obligation on his part to use her services, cutting down another objection he might have to the meeting. This is not the only possible cover letter Jennifer could have written, but this one will likely go a long way toward getting her in the door for a meeting.

Notice that Jennifer closes her cover letter by saying she will be calling the prospect to set up a meeting. That's smart. You can't expect the prospect—no matter how much free advice you're offering or how impressed they are with you—to call you. If you call first, you also have the advantage of being primed for the call and may be better able to steer the conversation toward the meeting you want.

There are a lot of things the material you send to the prospect can be, but there's one thing it should never be: a resume. A resume may be fine if you're looking for a position as a permanent employee, but not if you're trying to land assignments as an independent contractor. A resume focuses on who you are and sells you, but as an independent contractor you're selling *solutions.* You're not trying to convince the prospect that you're a well-qualified person; instead, you're trying to convince the prospect of your problem-solving abilities.

Of course, you'll have to provide the prospect with some information on your background, but that isn't enough. As we said in the last chapter, you have to relate your attributes in some way to solutions to the client's problems. (If you went through the exercise in the last chapter of defining just what type of "product" you are, then you already know how to relate your attributes to the client's needs.) A conventional resume leaves too many questions in the prospect's mind as to what you actually *do.*

Besides, you promised the prospect some information that would benefit them in some way. How are you going to do that with just a resume? The answer is that you can't. The written material you send to prospects has to be more like a direct mail sales pitch than a conventional resume.

> **Newsletters**
>
> ***You might consider a monthly or quarterly newsletter as a client "freebie." It's inexpensive to produce using desktop publishing, and is a great way to keep your name in the front of a client's mind. The newsletter can provide helpful hints, talk about ongoing work, chat about industry happenings, discuss upcoming relevant conferences or seminars, etc.***

Keep in mind that many people won't have the time, patience, or interest to read lengthy written material you might send them, even if it promises to help them in some way. For example, when was the last time you read an owner's manual for some product all the way through? We've faced that problem when producing such manuals for our clients. When people buy something new, they want to unpack it and get it running, not sit down and read a manual.

Our answer to this situation was a two-pronged approach: a quick summary of the main points up front in the manual followed by more detailed information further back.

This same approach can work well in material sent to prospective clients, with the "quick summary" taking the form of typical problems or questions the prospect might face, followed immediately by solutions. Let's suppose that the service you're offering is the development of custom multimedia-based software for personal computers. Your "quick summary" might be something like the piece shown on page 80.

Note carefully the scope of these questions and the information provided. Each is a legitimate question the prospect might have. The answers are accurate and useful instead of self-serving fluff. *Yet the information given does not in any way substitute for or diminish the need for the contractor's services.* The "free advice" quickly establishes the credibility of the independent contractor as an expert in multimedia and shows the prospect that multimedia might well be a feasible solution to some company needs. But there's only one way the client can know for sure, and that's to get the independent contractor in for a meeting! This is the essence of your written material—give enough of your expertise to provide some real value to the recipient while leaving them hungry for more.

Your qualifications and attributes must be linked to the questions and solutions you've raised. Rather than put this information in paragraph or text form, it's best to put it in short "bullet" form like this:

- ***I am a certified Windows software developer. My thorough knowledge of this operating system allows me to exploit all multimedia features of Windows.***

- ***I have a thorough knowledge of the C programming language. I can write special routines to accomplish functions not provided in multimedia development software.***

- ***I am skilled in the use of multimedia development tools for Windows and Macintosh computers. In addition, I have the necessary conversion software to transfer multimedia applications between Macintosh and Windows PCs.***

Real Answers to Real Questions Involving Multimedia

There's a lot of excitement about multimedia, as it's clearly the wave of the future in personal computing. However, there's also a lot of misinformation and hype about multimedia. As independent multimedia developers, we're not "locked into" any hardware or software platforms. Instead, we choose the mix of technologies that promise the quickest, most reliable, and least expensive answer to your unique requirements. You may have questions about the use of multimedia technology in your company. Here are some of the more common questions we've received from our previous clients; perhaps you have similar questions:

❖ ***Which computer system is best for multimedia: a Macintosh or Windows PC?*** The Macintosh is the most "multimedia-ready" PC, as the necessary video and sound capabilities are built into the Macintosh. However, recent versions of Windows (3.1 and later) give superb support for multimedia. A Windows PC with all necessary hardware and software to support multimedia is less expensive than a comparable Macintosh, and more "off-the-shelf" multimedia packages are available for Windows PCs. Windows PCs are rapidly achieving acceptance as multimedia development platforms.

❖ ***Is a CD-ROM drive always necessary for multimedia applications?*** No. A CD drive is useful if you want to run several different multimedia packages on the same computer or have a very large application. However, it is also possible to install a smaller multimedia application on a hard disk drive. If you are only going to be using one multimedia package—such as a training course or a product demonstration—then putting it on the hard drive can be a cost-effective approach.

❖ ***Can multimedia be used with laptop computers?*** Yes, indeed! Whether a PC is a laptop or desktop model has nothing to do with its ability to be used for multimedia applications. Instead, the key is whether the PC meets certain hardware and software requirements. These requirements include using Windows 3.1 or higher operating system, a 486 or higher processor, a super VGA color monitor, at least 8 megabytes of RAM, a SoundBlaster or equivalent sound card, and a hard drive large enough to hold the finished application. Numerous laptop PCs meet these criteria, while some desktop PCs don't.

Since you're selling solutions to client problems, the "meat" of a traditional resume—your previous jobs and education—is saved for last and then is usually given only as a quick summary. The only real reason you include this information is because some clients will be curious about it. Something like the following will usually do:

> ***Before offering my services independently to clients, I held permanent software development positions with Microsoft, Asymetrix, and Aldus. I received a B.S. in computer science in 1988 from the University of Washington. I received certification from Microsoft as a Windows developer in 1991 and LAN development certification from Novell in 1992.***

This format for written material works surprisingly well for a wide range of services. Are you planning to offer services for those interested in starting their own retail stores or restaurants? Your "free information" for them could deal with subjects like business licenses, ordering through wholesalers and distributors, and state labor laws. Are your services of interest to those in import/export businesses? They might welcome hints on how to deal with customs regulations, save money on shipping, or speed up payment from overseas. Are you a former teacher wanting to do corporate training? Then give free advice on training methods for adults, whether classes should be held at the company's facilities or away, whether classes should be held during work hours or later, etc.

You may want to get help from other independent contractors in preparing your written materials unless your writing and design skills are strong. Nothing turns off prospects faster than misspellings, grammatical errors, and material that's difficult to read because of a weird typeface. You don't have to pay top dollar for a slick, full color brochure, but you do need something that looks like it came from a serious professional.

We have to add at this point that the written material format discussed in this section won't be best for everyone. Artists' materials are normally almost entirely visual, with very little reference to their education or prior employment (those items are irrelevant; either prospects like the art samples or they don't). If your services are more research-oriented (especially in the sciences), then you'll need to give your academic credentials and previous experience (such as publications in professional journals) more display earlier in your material. The same applies if attainment of certain degrees or experience is a key component of your services. For example, your credibility as an independent contractor specializing in integrating IBM computer equipment with non-IBM equipment would certainly be enhanced by experience as an IBM employee who performed those tasks. In this case, stressing your previous employment in your written material would be a good move. Don't be afraid to vary from the format we suggest if your circumstances suggest doing so.

Handling Resistance

So far in this chapter we've been operating under the assumption that things will go smoothly when you call to get a name or make your initial and follow-up calls to a prospect. As you might suspect, things don't go nearly so well in the real world. You'll run across people who won't give out names, won't take your calls, or won't meet with you.

We've already mentioned the only two ways to handle such situations. The first way is to treat everyone you encounter at the prospect company with professionalism and respect. It does no good to get angry at secretaries or assistants who won't put you through to their bosses. If you manage to get through to a prospect on your third phone call, it's stupid to start off by chastising him or her for not returning your previous two calls. And if a prospect says your services are not needed, you're not going to change his or her mind by saying how shortsighted they're being for not wanting to meet you or that you know their businesses better than they do. You might think we're stating the obvious here, but—we swear—we have received calls exactly like these at HighText in the past. If you can't convert a prospect into an actual client, leave them feeling friendly or at least neutral toward you. You can always go back to them six months or a year later. People tend to remember the fools and boors they encounter—they also swap horror stories about them with people they know in other companies—and the last thing you want to do is convert a prospect into an enemy.

The second way to handle resistance is the previously mentioned "rule of three." One reason why this rule works is that it walks the thin line between being persistent and being a pest. If there is possibly a real need for your services at a company, then they will almost certainly make some positive expression of interest (such as accepting your phone call) by no later than your third attempt. If they don't, then the odds are very good they legitimately have no need for your services. Moreover, making more than three calls doesn't communicate your interest and enthusiasm; it communicates your desperation for work and inability to take a hint.

Remember that not all resistance you encounter means the prospect is not interested in your services under any circumstances. A "no" might just mean "I'm too busy to talk to you right now," "I have to talk this over with my boss first," "I need to get more information before I can make a decision," or "I don't know what you're talking about and I don't want to look stupid." Sometimes a "no" is just a holding action or somebody's reflexive response to anything new. While you have to outwardly show respect to the first or second "no" you hear from a prospect, you don't have to take it as the prospect's actual response. If they say "no" a third time, then they're probably serious.

The key to handling resistance is to accept and seemingly agree with the response. Don't overtly disagree and challenge the prospect; you'll only force them to defend their response or hang up on you. Let's suppose a prospect says something like, "We've never had good results with people from outside this company." Rather than trying to convince the prospect that he or she is wrong, say something like "I've heard that from other companies. What particular difficulties have you had?" If a prospect tells you "We just have no need for someone who does what you do here," respond with "It seems I received some erroneous information about your company, because I thought there was a good fit between my background and your company's business. Have you ever used the services of outside people in related fields?"

By responding to resistance with a nonaggressive question, you at least keep the prospect talking and get closer to the reasoning behind their response. For example, let's suppose that a prospect responds to your question about difficulties with other contractors by saying how such contractors have repeatedly missed deadlines. Your first impulse might be to assure the prospect you'd never do such a thing. Instead, *agree* with the prospect. Say something like "I've heard the same thing from other companies. I don't like those people either since I get tarred with the same brush. That hurts, because I take special pains to be conservative in the deadlines I set, and my other clients can tell you the importance I place on delivering on my commitments . . ."

Don't disagree with the prospect, even if you do. Disagreeing at this early stage will only shut down the line of communication you've opened. If you can't honestly agree with the prospect's objections or problems, at least keep the conversation going by using such nonjudgmental phrases as, "I'd like to know more; could you please elaborate on that?"

However, in some cases the first "no" might really mean "no" and you have to be discerning enough to recognize that. At HighText, we do no military ("mil spec") or government technical manual projects at all. However, we constantly get queries from independent technical writers whose backgrounds are exclusively in military and government work. When we tell such people "no," we really mean it the first time. In such cases, you might ask if the prospect knows of other companies that might need your services or if the prospect anticipates any change in their situation in the near future. But be prepared in some cases to realize there is no need for your services at that company. Politely say goodbye and call the next prospect on your list.

What happens if you have the name of a prospect, yet he or she never returns your calls? If this person's calls are answered by a receptionist or secretary, ask when it would be best to call. Some people do not accept outside calls at certain times of the day or when they need to work without interruption. If you treat the receptionist or secretary with courtesy and respect, they may give you hints on how to get your call through or at least returned. A riskier strategy we mentioned previously is to try calling before or after normal working hours at the company. The receptionist or secretary is likely not to be there, but the person you want to reach might be in their office. If the prospect doesn't return your third call (either via a receptionist or voice mail), at least send them your written material. In your cover letter, just mention that "Here is some material on the subject I was calling you about." Then try another call about a week later.

Keep track of those prospects who turn you down and note the reasons why. Things change. Six months or a year later, someone more receptive to using your services may be at that company or the company might enter a new area where you could help. Making additional approaches to "turn-down" prospects at decently spaced intervals certainly can't hurt, and many independent contractors use such re-contacts as opportunities to experiment with new phone techniques or written materials.

Chapter 4

Your Foot Is In The Door—Now What?

Once you've managed to arrange a meeting with a prospect, your next goal is simple. You want to end the meeting by saying something like this:

I really appreciate your taking the time to meet with me and allowing me to find out more about your company. You've given me a lot to think over. Let me suggest this: I'll prepare an analysis of your situation and include my recommendations on how to address the issues we've discussed. I'll also include a schedule of how long it would take to perform all tasks and a breakdown of the costs involved. Of course, this doesn't obligate you in any way. If you agree with my approach, I can start work upon your acceptance of my proposal.

If you're really good at this (or really lucky), the prospect might ask you directly to prepare a proposal. At any rate, you must produce proposals in order to turn prospects into clients. You want to learn enough about the prospect's problems during the meeting so you can submit a proposal that's likely to be accepted. The way to do that is to get prospects to talk about themselves. The more your prospect talks—and the more you listen—the better you will be able to understand their problems and develop solutions you can propose.

Prospects will be using the initial meeting to size you up and form a judgment as to whether you can help them. But the meeting with the prospect isn't a one-way street; you're evaluating the prospect just as much as they're evaluating you. You want to learn enough about the prospect to know when you don't want to submit a proposal to them.

Clearly, a lot is riding on your meeting with a prospect. Let's examine how to go about conducting it right.

The Two Stages Of Any Initial Meeting

There are two clearly distinct stages to any initial meeting. The first is brief, usually lasting only about five minutes. In this first stage, the prospect is forming some judgments about you that will likely be the determining factors in whether or not you get an assignment. In this stage, the prospect is not looking to determine how well qualified you are or what your background is; instead, the prospect is looking for something more elusive, namely how *comfortable* they feel with you. Keep in mind that many of the people you meet with will be taking a big risk in giving an assignment to an outsider. If you fail to do what you promise in a satisfactory manner, you might cost them their job or their business.

The first priority of a prospect will be whether they feel they can trust you with an assignment. A lot of social science research confirms what common sense tells us: first impressions count for a lot and are often decisive. During the first phase, the prospect will form a judgment whether he or she trusts you based upon factors—your demeanor, your appearance, your conversational ability, etc.—that have nothing to do with your skills or experience. It's true that negative first impressions can be reversed over time. But if you're an independent contractor, the prospect won't give you the time to overcome a bad first impression.

The second stage involves your evaluation of the prospect. As we've noted, you must find out enough about the company and the problems it faces to allow you to submit a proposal for their consideration. But you are also trying to find out more about the dynamics of the company and how that might affect any work you do for them. For example, some companies will have problems because they are incompetently managed or undercapitalized. Even if you have the skills to help the company and they want you to do so, you might decline to submit a proposal if the company's fundamental problems are beyond your ability to solve.

You must execute both stages of the initial meeting well. If you don't handle the first stage well, you'll never get an assignment. And if you don't do your job in the second stage, you could wind up with an assignment straight out of hell.

Doing Your Homework

When you first meet with a prospect, you want to be armed with every bit of information possible about that company. You want to know who its key executives are, the number of employees it has, its main products, and the types of customers it serves. Having such information lets you and the prospect make efficient use of the second stage of the meeting. And it never fails to impress prospects when you walk in loaded with facts and figures about them. You can find such data in the same industry and regional directories you used to compile your list of prospects. Other good sources for pre-meeting research include back issues of trade and industry publications (especially those with an annual index), advertisements for the company, back issues of the business section of your local newspaper or business papers like the *Wall Street Journal* (these will be on microfilm at your library), and company catalogs and other sales materials. You can often pick up the latter by stopping by the company ahead of time and asking the receptionist for them.

Speaking of dropping by the company in advance, it's a good idea to take a look at the company's facilities before your visit. The type of building the company is located in and the neighborhood around it can tell you a lot about their self-image and—sometimes, but not always—their finances.

As for finances, it's important to keep in mind that you have to get paid for the work you do. If you're dealing with a established company, you're usually on safe ground. However, newer and smaller companies can pose some risk, especially in their start-up phases. Such companies also happen to be among the biggest users of independent contractors. In your pre-meeting research, you might discover that the prospect had some past cash problems or some recent losses. If you find such information, that doesn't necessarily mean you should cancel the meeting or raise the subject in the initial meeting. However, you will need to include some way to protect yourself, such as a substantial down payment or staggered payments through the assignment, in any proposal you might make.

One way to determine a company's payment history is by purchasing a credit report from such agencies as Dun and Bradstreet or TRW. This can be expensive, however ($50 to $100), and the information in such reports is not always complete or current. It's usually not cost-effective to get a formal credit report in advance, but you may want to do so after your initial meeting if you have some qualms about the prospect's financial status. By the way, don't be fooled into thinking that a company is prosperous simply because it has expensively decorated offices. Perhaps the most beautiful and impressively decorated office building we ever visited was rented by a company—not a Fortune 100 company, by the way—that had just lost over $35,000,000 in the previous year! By the same token, a lot of well-funded new companies put most of their cash into product development and marketing and furnish their facilities with used furniture.

You should also prepare a few stock answers for some questions that permanent employees love to ask independent contractors. Some corporate employees will see you as something exotic—*You walked away from a position with MegaCorp! I can't believe it!*—and you're likely to hear questions like these:

- **Why did you decide to go out on your own?**
- **How's business?**
- **Can you really make any money doing what you're doing?**
- **Aren't you afraid of not being able to find any work?**
- **How long have you been doing this?**

Such questions are actually great opportunities to start selling yourself and your skills. You could answer the first question with something like "What I really enjoy is working directly on a problem and solving it. It's also what I happen to do best. However, I had reached the point in my career where I was spending more time with paperwork and supervising others rather than actually working on problems. I've been a lot happier, and I think I'm doing better work, since I started offering services on an independent contract basis to companies like yours." If you're asked the last question listed above early in your contracting career, truthfully answer that you're new to contracting and instead stress your experience as a permanent employee.

Finally, you need to have an agenda in mind for the meeting—after all, you asked for it and it's up to you to provide direction. The prospect obviously had some reason for agreeing to meet with you (if only curiosity) but they may not have any plan for the meeting. If it becomes clear the prospect has their own agenda for the meeting, defer to it. But don't count on the prospect having anything planned, so be ready to steer the meeting in the direction you want it to go.

Know Your Competition!

As part of your overall preparation for selling yourself to prospective clients, be sure you understand your competition. Make it your business to know who else offers comparable services in your business area and find out as much as you can about them. We recently received a mailing from an independent contractor offering book production services, and she clearly and specifically addressed in her brochure what makes her services different from her competitors in southern California. We were impressed, and she will probably end up getting some business from us! Be prepared to offer a prospective client detailed reasons why he or she should use you instead of one of your competitors.

The materials you bring to the meeting depend on the nature of your services and the examples you want to show of your prior work; you may want to leave some of these with the prospect. Take additional copies of the material you sent to the prospect and business cards in case there are additional people at the meeting. One thing you should always bring along is a notebook and pen to take notes of the meeting; you'll need these to prepare your proposal. It's best to avoid using a tape recorder in your initial meeting, since many people "freeze" when they know they're being recorded. (However, a tape recorder can be a big help in subsequent meetings if the client is agreeable.)

Your Image

Your mother was right—first impressions do count for a lot. If you're going to successfully make your way through the first stage of your meeting, you're going to have to pay attention to the sort of image you project to others.

Making people feel confident about placing a lot of responsibility in your hands starts with the obvious things like clothing, grooming, and punctuality. Unless you're certain you can get away with something else, always wear conservative business attire to your initial meeting with a client. People assume that if you dress like a slob that you'll put the same care into your work. And if you can't be on time for your initial meeting, why should the prospect feel confident you'll be on time for anything else? It's a good idea to show up early for your first meeting. This lets you get a feel for the general "atmosphere" of the company—do the employees seem cheerful? Energetic? Glum? Or even fearful? You may be able to engage the receptionist or others in conversation and pick up more insights.

Your total demeanor for the initial meeting must be friendly, professional, and empathetic toward the prospect. Think back to the doctors and dentists you've known and which ones you felt best about. The ones you remember most favorably are likely not always the smartest or those with the most impressive degrees on the wall. Instead, you probably most favorably remember those who seemed to really listen to what you had to say, seemed interested in your problems, clearly explained the procedures you were going to undergo, and projected a confidence that something could be done about the problem that was bothering you. Those are the same feelings you must instill in the prospect. If your natural personality is morose or argumentative, then you'd better learn how to turn on the charm around prospects. Likewise, being nervous, distant, or cold won't win over many prospects. This doesn't mean you should be a backslapper, or exuberant about anything and everything (wouldn't you feel a little nervous seeing a dentist who acted like a lodge brother at the annual convention?). All you have to do is act in a friendly, quietly confident manner. If you feel your personality might be lacking in some areas, then attending something like a Dale Carnegie course or other training in interpersonal skills would probably be a worthwhile investment.

Projecting the right image starts the minute you walk in the door at the prospect company. Always greet the person you're meeting with first. When someone who might be the person you're meeting enters the reception area, stand and say something like "Bob Williams? Hi, I'm Jennifer Roberts . . ." If the person you've greeted is not Bob Williams, don't worry; the person will let you know by the time you've uttered those words and your effort won't be held against you. (In fact, the person you greeted might remark to Bob Williams how friendly you were.)

If the person is indeed Bob Williams, step forward to meet him.

Give a firm handshake, smile, and make eye contact. (The handshake shouldn't be as firm as Arnold Schwarzenegger's, but neither should it be like handing the prospect a dead fish.) If possible, try to be the first to say something after the initial greetings, and try to make that something a question the prospect will be proud to answer. Don't ask something negative like, "Is the traffic around here this bad all the time?" (Don't laugh; we've been asked that!) Instead, ask "The design of this building is really interesting; whose idea was it?" or "That's a creative corporate logo; who designed it?" There's no need to be fawning in your questions; just pick something that most people would positively remark about (this is another good reason to show up early—so you can pick something to ask about). As you are escorted to the meeting room, try to keep the conversation going with further questions or open-ended comments ("I was looking through your catalog, and you have some really fascinating new products available") that invite a response. Remember that the person who met you may not be much of a conversationalist or may be just plain shy; it's up to you to keep things moving.

> ***"In my first meetings with prospective clients, I try hard to counteract the typical image people have of an 'artist'—temperamental, unrealistic, off-the-wall, and so forth. I try to impress upon them that I respect their time and their hard-earned money. I've found that the direct approach really works best for me. I don't beat around the bush. I just ask them plainly what I want to know—exactly what it is they're looking for, what it would take for them to decide to give the work to me. And if I end up not getting the job, I ask why."***
>
> **Brian McMurdo, graphics artist, Valley Center, California**

You might be met by the prospect's assistant or secretary. If that happens, be friendly and ask questions; treat the assistant just as you would treat the prospect. The prospect might well ask the assistant to give his or her opinion of you, and that opinion may be a decisive factor in the prospect's reaction to you. Always assume that everyone you meet in the prospect company has some influence on whether you get an assignment. Treat everyone with courtesy and professional respect.

Meetings With More Than One Person

You might find other people waiting on you in the meeting room. Even if you've talked to just one person at the prospect company, it's not uncommon for other people to be present at the meeting. Don't be surprised if this happens and take it in stride. Introduce yourself to each person with a smile, eye contact, and a handshake. Then offer your business card. Ask if they would like a copy of your written material. If other people join the meeting after it's begun, stand to greet them and repeat the process. If the other people you meet don't have business cards, quickly make a note of their name and position.

If more than one person is sitting in on the meeting, one of the key things you need to determine is who the ultimate decisionmaker is. Sometimes this will be obvious by job title (if there's only one vice president in the meeting, it's a good bet he or she has the final approval). In other cases, it won't be obvious. In such cases, pay careful attention to the dynamics of the people you're meeting with. Eventually you'll notice that the others seem to be deferring to and avoiding confrontations with one person. That person is almost certainly the decisionmaker, even if another person (usually the one you contacted) is running the meeting. Regardless of who is running the meeting, treat everyone at the meeting as if they were the decisionmaker . . . because you might have misjudged the situation and one of the "lesser lights" might actually be the one who decides if you get the assignment.

One problem with meetings that have several people attending is that they tend to be less informative than meetings with just one person. While no person you've just met is likely to confide company secrets to you during the initial meeting (although it has happened to us), many people are reluctant to admit that anything at the company is less than perfect if other employees are around. In such situations, you may have to ask essentially the same question in several different ways before you get the desired information. You may not be able to get enough information from the various participants to formulate a proposal, especially if the task or problem crosses department lines (for example, a manufacturing company that needs to better control its raw material inventory may have troubles in both its production and purchasing departments). You may need to get additional information later by telephone or in separate meetings with individuals attending the initial meeting. If you sense this is true at the end of your first meeting, ask if you can later follow up on some of the points raised by telephone or suggest another meeting with some of the participants.

If you do wind up in a meeting with several people, take it as an early warning sign that the company may be managed by committee and it may be difficult to get a decision made on any proposal you submit. The probability of this being true seems to increase in direct proportion to the number of people in the meeting. If you're facing ten people, you have a real problem on your hands! And don't be surprised if such companies have some deep-rooted management problems caused by their tendency to analyze things to death and an inability to make a quick, definitive decision.

The Two "Don'ts"

Two questions are likely to be asked during the initial meeting (you may even be asked them during your phone call to the prospect). These questions may pop up any time during the meeting, so you have to be prepared to field them from the start. The questions seem harmless and logical, and it's a big temptation to give answers. But if you do so, you'll shoot yourself in both feet.

The first question goes like this:

How much would you charge to do this?

This seems reasonable enough; most of us want to know what the price of something will be before we seriously consider buying it. You'll eventually have to name a price before the prospect will consider your proposal. But the initial meeting is not the place to discuss price, and you'll regret it if you do.

Why? For one thing, you know nothing at this point about what the prospect needs to have you do, nor do you know enough to determine how long it will take you. Even if you wanted to answer the question, you simply don't have enough knowledge to give an accurate, meaningful answer. If you try to quote a price high enough so that you won't lose money regardless of what you have to do to complete the assignment, you may quote a fee beyond what the prospect wants to pay. That's bad enough, but the alternative—quoting a price so low that you actually lose money on the project—is even worse.

The correct way to answer this question is something like this: "That's a key question for you, and I wish I could give you a straight answer now. However, I need to learn more about your company and any tasks you may want me to perform before I can give you an answer that's valid. As soon as I have a good understanding of your situation, I'll let you know my fees, and I'll put them into writing."

The prospect may try to ask the same question from a different angle, inquiring about your hourly rate. Even if you have an hourly rate in mind, don't take the bait. Instead, say that you sometimes charge by the hour or by the task. (We'll discuss pricing later, but for now just say something like this even if you only charge either by the hour or by the task.) Say that the exact amount you charge varies with the complexity and difficulty of the assignment, with other factors, such as schedule, taken into account. Tell the prospect that they are probably more interested in knowing the total amount the assignment will cost rather than the hourly or per-task rate, and that you'll give them a written quote as soon as you get enough information about the project to make an accurate quote.

Whatever you do, stick by your guns and don't name your price in the initial meeting. Almost all prospects will understand the merits of your position and not press you further. If you do run into a prospect who insists on a definite quote in the initial meeting, then you're dealing with someone who places more importance on how much something costs than upon the quality of the work or other factors. Odds are that you don't want a company like that for a client, because they will always be looking for a way to save a few dollars, and how much they pay you will likely be one of the first places they'll look. Moreover, if money is that tight for them then you should legitimately wonder how soon—or even if—you will get paid.

The second question to look out for is this one:

How would you do this?

This question is dangerous for several reasons. One is that the prospect is, in effect, asking you to work for free. If the answers to the prospect's problems are obvious, then a full and accurate answer will give the prospect a good start on how to solve their problems without you. A full answer to this question also can hit you in your bank account. If you can really answer this question to the prospect's satisfaction in the initial meeting, then the prospect isn't likely to pay you much even if you do get an assignment. (If you can come up with the answer so easily, then what you're doing can't be all that difficult!) Another danger is that you can undermine a prospect's confidence in you if you attempt to answer this question. The prospect may fully understand that the company has some severe problems, and maybe their own attempts to resolve those problems have failed. If you come in and after a few minutes of conversation claim to have all the answers, then you may be perceived as a glib B.S. artist instead of a skilled professional.

The way to handle this question is not to give any concrete answers, even if the solutions to the problems are obvious to you. Instead, describe the *process* you would use to go about finding a solution to the problems. A good response might be "I honestly don't know how to do that yet. I need to learn more about your company and the situation you're in before I can make some concrete recommendations for action. However, I can tell you how I would go about learning what your problems are and developing ideas on how to solve them. I'd start by . . ." You could then describe how you'd like to meet with certain types of people, the records and data you'd like to examine, what you'd like to observe—whatever you would normally do to determine exactly what the problems are and how they can be resolved. You're not really giving anything away, since you'll describe the process in generalities and any solution depends upon making use of your skills and judgment.

When prospects ask these two questions, they're not being sneaky or devious. In almost all cases, they don't appreciate the magnitude of what they're asking. You must be aware of what they're asking, however, and refuse to answer either question in the first meeting.

The Meeting Itself: Making Them Talk

As noted earlier, the first stage of the meeting process—where you and the prospect size each other up and start to build trust in each other—is quick, lasting only a few minutes. Typically the conversation will be about the weather or some event in the news instead of the reason for your visit. During this stage, let the prospect lead the conversation and decide when to end it and move on to talking business. You can do your part to indicate interest and establish a rapport by continuing to ask questions and make open-ended observations that invite a reply. If the prospect asks you a general question—"Who do you think is going to win the Super Bowl?"—answer and then turn it around so the prospect can also answer—"I like Dallas. And you?" This sort of interplay keeps the conversation going. Continue making eye contact with everyone you're meeting, smile, and project enthusiasm and energy. Eventually, the prospect will indicate when it's time to get down to business.

In that second stage, you start to move the meeting in the direction you want it to go. Regardless of whether one or several people are in the meeting, you must get them talking about the company and the situations it faces. If you're lucky, the prospect will already have some idea of what their problems or needs are and will share that with you. In other cases, the prospect may have nothing more concrete than a vague feeling that something is not quite right. In either case, there's only one way you're going to find out—you have to ask them questions and keep them talking.

You should take notes if at all possible, but it's always polite to ask first. Take out your notebook and pen and ask, "Do you mind if I take notes about our conversation?" The answer will inevitably be yes, and your courtesy in asking will be appreciated.

Your questioning has to be gentle and inoffensive. If you start peppering the prospect with questions as if you're conducting an oral exam or a criminal cross-examination, you'll probably turn off the prospect. No one likes to be pressured or put on the spot, and a too-aggressive style of questioning will do just that. The best way to avoid doing this is to start out with the sort of general questions anyone might ask about their business and gradually move to more targeted questions as the conversation progresses.

Suppose that you're offering services as an industrial designer. Instead of immediately quizzing the prospect with questions about their industrial design needs, start out with some overall observations about their company and its products. "I notice that there seems to be three distinct looks to your company's products. Am I right?" will probably be enough to trigger a response that will lead to a discussion of the company's design history and philosophy. You can then direct the conversation through your questions to the company's most recent design efforts, which will let you determine whether they're happy with their current designs and what changes they might be looking for.

By this point, you may have established a solid-enough foundation with the client that you can ask something as straightforward as, "Are you happy with the most recent design as it is or are you looking for some refinements?" A specific question like this early in the meeting might produce a defensive response like, "Oh, we're very happy with our designs." But if you save it until the latter stages of the meeting, you may well get a more honest answer like "Well, we were when it was introduced but now we feel the design looks a little dated, especially compared to our competition." When you get an answer like that, you've struck pure gold. The prospect is about to tell you what they want you to do for them.

Your response shouldn't be "So what would you like changed about the design?" The prospect might not be able to articulate exactly what is wrong with the current design; they may only know something about it doesn't please them. A general query about what they don't like might just frustrate the client or produce defensive behavior ("Well, it's really not that bad . . ."). The best way to quiz the client is by isolating possible problem areas one by one. You can start off your questioning with "Let me get a better understanding of what you mean . . ." and then ask things like the following:

- "How about the design of the handles? Do those seem okay to you?"
- "Are you bothered more by the shortages you have in your inventory situation or by the excesses?"
- "How big a percentage of people who make reservations never show up?"
- "Is your employee turnover higher than you'd like?"

The examples above—in order, they're from the fields of industrial design, inventory control, restaurant management, and personnel management—illustrate how this questioning technique works no matter what services you're offering. Keep refining and narrowing your questions according to the prospect's responses. For example, suppose the prospect says the handle design is a problem. Your next series of question would try to isolate exactly what is wrong with the handles. Are they too small? Hard to grip? Prone to breakage? Such questions are helpful to those prospects who haven't identified exactly just what about their current situation bothers them.

Once the prospect has identified the problems for you, try to determine what, if anything, the prospect has tried to do about the problem. A query like "Have you tried to get the handles redesigned since they were introduced?" may be just the trigger for the prospect to discuss why the company has been unable to solve the problem on its own. You can learn about solutions that have been tried but failed and perhaps why they failed. (If they don't know why they failed, that's a problem they probably want you to solve!) This sort of information is invaluable in formulating your proposal.

During the course of the meeting, the prospect may say something that you disagree with, perhaps strongly so. However, openly disagreeing with the prospect is a good way to bring the dialogue to a screeching halt; the prospect will feel compelled to defend their position and will view you as an antagonist rather than a potential ally. Moreover, disagreeing with a prospect won't get you any closer to what you really want to know, namely why the prospect feels that way. Instead of saying "That's wrong," try "That's a little different from my experience. Could you explain why you feel that way?" instead. Even if the prospect invites you to disagree, don't take the bait. If the prospect replies, "So, you disagree with that?" you can say, "Well, it is different from what I've encountered in the past, but there may be a good reason for that. Could you tell me more?"

You should also avoid directly agreeing with the prospect unless it's a truism ("Yes, customers are the most important thing to any business!") or related directly to getting the assignment ("Yes, I'd be happy to prepare a proposal for you.") Otherwise, just react with neutral phrases like "I understand." One reason you want to do this is that you're in an information-gathering mode. At this point, you have no firm basis for agreeing or disagreeing with the prospect. The prospect will almost certainly be telling you the truth as he or she sees it, but the objective truth of the situation could be vastly different. If you indicate agreement, you may lock yourself into a position that you later find is not supported by the facts.

Another problem is that you may unduly influence the

prospect in a certain direction. If you overtly approve of what she is saying, she may unconsciously seek more expressions of approval from you instead of dealing with subjects that might be unpleasant. It's instead better to give subtle, nonjudgmental verbal and visual cues encouraging them to continue. A quick "I see" will do wonders to keep the prospect talking, as will smiles and nods of your head. When you make a comment or ask a question, be quick and to the point about it; don't break the prospect's train of thought.

As the discussion of various topics seems to end, it can be useful to summarize what you've heard and get the prospect to verify it. Start with a phrase like "Let me make sure I've understood you correctly . . ." or "In other words . . ." and then give the main points of what you heard. The prospect will often be prompted by your summary to add a point or two or explain something further.

When the meeting seems to be drawing to its natural close, or the prospect indicates that the time is up, offer to submit a proposal. If you are meeting with more than one person, ask who should receive the proposal. Give the prospect a date by which they can expect to receive it, and ask if you can telephone in the interim should you have any questions or need clarification of any points.

In some cases, the prospect may offer you an assignment at the end of the meeting. This is especially true if you offer a service (graphic arts, technical writing, computer programming, etc.) where a "trial run" with a small project is possible. If you want the company as a client, accept their offer.

There's no big secret to holding useful meetings with prospects. Be positive, friendly, and upbeat. Start asking general questions and then gradually move to more specific questions. Listen, listen, and listen some more and take notes. Don't disagree with the prospect; instead, ask them why they feel the way they do. Summarize the points the prospect makes and offer to make a proposal. And don't blurt out the answer to the prospect's problem (even if the answer is obvious) and don't discuss a price. With a little experience, these things will become second nature.

Reading The Client

Back in the 1970s, there were several books written about "body language." The theory behind these books was that people indicate their emotional state by how they position their bodies and use their arms, legs, etc. A lot of those books were junk, but two good ones were *How to Read a Person like a Book* by Nierenbery and Calero and *Body Language* by Fast. You might want to locate these at the library.

You don't have to read such books to know that certain movements and body placements have meaning. Here are some of the more common signs to watch for:

- If the prospect is relaxed, making direct eye contact, and has no arms across the chest, then he or she is receptive to what you're saying and is probably satisfied with the way the meeting is going.
- If the prospect is leaning toward you or sitting on the edge of the chair and is making steady eye contact with you, he or she is probably very interested in what you're saying.
- If the prospect leans back with hands clasped behind the head, strokes the chin or neck, or doodles and watches the results, then the prospect is thinking and evaluating.
- If the prospect has arms folded across the chest, turns his or her body away from you, wrinkles the brow, and/or avoids eye contact, then the prospect is defensive about what is going on or feels threatened.

None of these indicators is 100% reliable; some people are good actors while others who look relaxed are simply not interested. However, it is important that you look for these and other cues as to how the prospect feels and react accordingly. For example, you may notice that a certain subject makes the prospect break off eye contact and shift uncomfortably in the chair. Depending on your "feel" as to how the meeting is going, you may want to ask the prospect if that subject is a problem or wait to address that in your written proposal.

Meetings From Hell

So far, we've assumed the people you will meet will be civil, rational individuals. That is usually the case, but every so often you'll come across real clunkers. We've met with people who start off the meeting by complaining about what a lousy place they work for or start discussing their personal problems. In other cases, the prospect adopts an immediately aggressive posture or seems completely uninterested in talking, no matter how many questions you ask. How do you handle these situations?

In many such cases, as with people who complain about their employer or start discussing personal issues, there is no real alternative but to recognize that the meeting will probably be a total waste of your time. Such behavior is so egregiously unprofessional that it raises a red flag to alert you that the company probably has some big problems if they keep such an individual on the payroll. The business judgment of such people—indeed, their common sense—is highly suspect. It's been our experience that trying to perform an assignment for such a person is exasperating; they ignore what's in the written agreement, don't provide the input they're supposed to, forget about their deadlines, expect you to hustle to make up for their failings, and you wind up calling them repeatedly asking where your check is when the assignment is completed.

So you should just bring such meetings to a close as rapidly as possible and get out of there, right? Not necessarily. For one thing, you can always use the meeting to try out some new questioning techniques that might be too risky to use with a better prospect. You can still learn some useful things during the meeting, as people who don't mind trashing their employer to a complete stranger usually don't mind discussing confidential details about their employer's operations. You might be able to put such information to use in an assignment for one of their competitors. And, in a few cases, you might actually be working with a far more competent individual during the assignment. So grin and bear it if you find out you're meeting with a dunce. Be polite, but be very careful about actually entering into any agreement with such people.

A more difficult situation is the prospect who refuses to communicate. You ask questions, show interest in whatever is said (no matter how briefly), and do all the things that work with other prospects. But, despite your best efforts, the prospect remains The Great Stone Face. After a while, you start to wonder why they agreed to the meeting in the first place. In such cases, there is really not much you can do other than to keep asking questions and trying to bring the prospect out of his or her shell. However, there is a high likelihood that the prospect was not planning to give you an assignment anyway when he or she agreed to the meeting. The motivations such people might have for scheduling the meeting could include getting price information (they might be wondering if they're paying their existing contractors too much, for example), curiosity (they might be interested in becoming contractors and wonder how you do it), or a simple lack of anything better to do.

If you're asked questions unrelated to the prospect's business (like how you go about finding clients) or are asked about pricing repeatedly, then the "prospect" is fishing for information instead of looking for someone to give an assignment to. Since you won't be able to get enough information to prepare a viable proposal if the prospect won't talk to you, you're not risking much if you take extreme measures to get the conversation going. If you're unable to get the prospect to talk to you despite repeated attempts, put your cards on the table and say something like "I'd really like to know more about your situation so I can determine if there is some way I can be of help to you, but I sense you're reluctant to talk to me. Is there some reason why?" If the prospect responds to that, at least you have some idea what's going on. If the prospect is still quiet, you haven't really lost anything.

Aggressive or hostile questioning by the prospect can have several causes. Sometimes it is a deliberate effort on the part of the prospect to see how you handle stressful situations. In many other cases, it is a defense mechanism by the prospect. The prospect may feel intimidated at having to ask for outside help or by your expertise. In a few cases, your very independence may trigger some unconscious resentment in the prospect. If the questioning is aggressive but not hostile ("I wonder if your background is a good match for what we need here"), take it in stride and continue to seek more information ("That's a concern of mine too, since I'm not suited for certain assignments. That's why I'd like to know more about . . .").

It's not uncommon to get an outright hostile question or two during an initial meeting, so be prepared to take those in stride as well. For example, a prospect might remark "We've never had good results with outside people we've used." Such a comment isn't directed at you personally; the prospect is just telling you how bad your competition apparently is! Instead of reacting angrily, you could respond with "I know there are too many incompetent independent contractors out there, and it makes my job more difficult. But please don't lump me in with them." This is the sort of strong but positive rebuttal that may even trigger an apology from the prospect.

However, you don't have to sit meekly through a meeting that turns into an inquisition. If the questions and comments from the prospect take on a decidedly nasty tone, don't be reluctant to cut the meeting short. A prospect with a chip on his or her shoulder is unlikely to give you an assignment, and, even if you do get an assignment, trying to please someone like that isn't worth whatever you're going to be paid. Politely but firmly put a halt to the meeting by saying something like "I appreciate your taking to time to meet with me, but I don't think this is getting us anywhere. I think it's best that I left now." Then leave! Permanent employees may have to put up with a certain amount of abuse, but you don't. Don't hesitate to cut short a futile meeting that turns nasty; you should use that time instead to locate a better prospect.

In meetings with more than one person, there may be some company political games and one-up-manship going on. Sometimes you'll witness a genuine argument among the members of the other side. In such cases, you can do little other than wait out these displays and try to get the meeting back on track with a question. If you observe a lot of such activity in the initial meeting, you probably have a good insight into a major cause of many of the company's problems!

After The Meeting

There are a lot of things you won't be able to put down in writing during the initial meeting. You will have developed impressions about the people you met, their relationships to each other and how they interact, the atmosphere and "feel" of the company, etc. These things can be crucial to your problem-solving approach, but obviously you can't risk recording them during the meeting. But do record such observations as soon as you can. We've done a lot of furious scribbling as soon as we've reached the safety of our cars in the parking lot!

Unless the meeting went so badly that you don't want to do any work for the prospect, always send a brief note of thanks to everyone who participated in the meeting. This little touch can be the thing that tilts a decision in your favor. Do this within 48 hours of your meeting. You should also review your notes in that same 48-hour period. If something isn't clear or you need additional information, call the person you will submit the proposal to. If you call within 48 hours, the meeting will still be fresh in the prospect's mind and they will have a better idea of what you're talking about. Delay longer, and the prospect may not remember what was or wasn't said in the meeting. Calling within 48 hours is also a good indication of your energy and interest.

At this point, you've finished the hardest parts of the independent contracting process. You've located a prospect, convinced the prospect to meet with you, and have held that meeting. Now comes the fun part: preparing and submitting a proposal.

Chapter 5

Getting Them To Say Yes—The Art Of Preparing And Presenting Proposals

You come back on a high from your initial meeting with the prospect. The meeting went well, and they were very receptive to your offer to submit a proposal for their consideration. You feel in your gut that your skills and their needs are a good match, and the work you plan to perform for them looks like it will be interesting as well as profitable. All that remains to be done is to get in front of a keyboard and start banging out a proposal for them, right?

Well . . . no. In fact, you're only half the way home. You haven't, as salespeople say, closed the deal. You've converted a prospect to a hot prospect, and that's an important step. But you won't make a dollar for your efforts unless you can get the prospect to accept your proposal.

The proposal is the last piece of sales material you submit to a prospect. While it's also a statement of what you plan to do and a contract for the work you propose, it's also your last attempt to convince the prospect to use your services. When you return from an initial meeting loaded with enthusiasm, it's tempting to sit down and get a proposal out as quickly as possible. But that's usually a bad move. The biggest reason for this is that no matter how well the initial meeting went or how long it lasted, *you probably don't have enough information to make a successful proposal.* A successful proposal isn't just one that the prospect accepts; it's also a proposal that works for you and the prospect and also lets you make a good return on your time and efforts. You need to contact the prospect again—usually by telephone—and get additional information before submitting your proposal. Having to contact the prospect again isn't a problem or a bother; it's a terrific opportunity. If you ask the right questions during the "re-contact," *the prospect will often tell you exactly what to put in your proposal to have it accepted.* In effect, the prospect will do the hardest part of proposal development for you!

By the way, a proposal isn't necessarily a lengthy, multipage document. For simple tasks or additional work done for a previous client, a proposal can be just a one-page letter. But every contracting assignment—even those where you've already received a verbal commitment from the prospect—should be governed by a written proposal of some kind. Putting the terms, conditions, and fees of the assignment into writing greatly reduces misunderstandings and other problems for you and the client. Regardless of its length, you need to go through the steps discussed in this chapter to develop a proposal that works for both of you.

Figuring Out What's Going On

Your first task after the initial meeting is to understand as best as you can what is going on in the prospect company. By "going on," we mean understanding the dynamics facing the company, such as its customers, competitors, how the company sees itself and its competitive environment, finances, interactions (such as rivalries) between company divisions and employees—anything that relates to possible problems and their solutions. This sounds calculating and maybe overly clinical—something like an autopsy—but it's necessary. Once you understand such things, you can then start determining what the problems are and how to solve them.

You won't have to do an exhaustive "situation analysis" for every potential assignment, of course. A lot of your potential assignments might be straightforward and unambiguous, but you'll still need to understand the prospect before you can make a proposal or realistically price your services. Let's suppose that you're an accountant and are preparing a proposal to handle accounting for a small business. This might seem a simple proposition, but how would it impact you if the employees of that company are sloppy about keeping receipts and other expense documentation? If it is your assessment of the situation that the prospect company has such problems, then you need to address them in your proposal and adjust your fees accordingly. (And if you have to chase down missing receipts as a regular part of your accounting practice for that company, then you need to charge the company for that.)

> ***"Be careful when you're asked to work for a fixed fee! I learned the hard way that I need to carefully pin down the amount of work the customer is asking for and the number of hours that it will take me. I've had some fixed-fee arrangements that were financial nightmares, where the customer kept adding on more and more tasks and didn't want to pay for them. I now ask for at least a partial prepayment if I have any doubts about a job."***
>
> Jim Penaligon, CPA, accounting contractor, San Diego, California

Here are some typical questions you should ask yourself regarding the prospect company:

✦ **What are the traits of the organization?** Is it an informal, hands-on, "shirt-sleeves" company or a formal, bureaucratic, "three-piece suit" one? Does it seem open to change or more concerned with preserving the status quo? Is the management style "hard" or "soft"? How well-managed does the organization seem to be?

✦ **What is the competitive environment the company faces?** Is it in an expanding, steady, or contracting market? Who are its main competitors? Who are its customers and what are their characteristics? What has been the recent history of the area of the company's main business? What are the future trends of the company's main business?

✦ **Who are the main players you will be dealing with and what are their characteristics?** The term "main players" includes all people who will have an impact on your ability to successfully complete an assignment, not just those you meet in the initial meeting or a designated contact person during the assignment. Do the people at the company appear cooperative? Hostile? Defensive? Indifferent? Competent? Does it seem that rivalries between different people and departments exist? Since you will have no ability to compel people to act in a certain way so you can complete a project, it's essential that you correctly evaluate how people in the company are likely to act.

✦ **How realistic is the organization's view of its situation?** During the initial interview, the company gave you information about the type of company it is, its products, and how it perceives its current situation. However, your perception of the facts may be very different. For example, people at a company may tell you they are in a growth market, while your conclusion might be that the market is stagnant and poised for sharp declines in the years ahead. You have to judge whether the company's view of its business, practices, and problems it faces is realistic or not.

✦ **What are the organization's values?** Some companies, such as investment bankers or funeral directors, place a great value upon presenting a dignified, sober face to the world. Other companies—advertising agencies or software developers, for example—often try to seem unconventional and even quirky. The organization's actual values may differ significantly from those it tries to communicate to the public. For example, the dignified investment banking firm may actually desire and encourage cutthroat, ruthless behavior. Any proposal you make has to be in agreement with the true underlying values of the organization.

✦ **Can the company afford your solutions—or even you?** Some organizations have champagne tastes but beer budgets. Others simply don't have the funds necessary to implement any solution (indeed, undercapitalization is probably the biggest problem most small businesses face). It does no good to propose assignments the company can't afford—even if your proposal is accepted, you still have to get paid!

Don't hesitate to do additional research, such as on the industry in which the prospect operates, to help you answer questions like these. However, you will probably leave the initial interview with some strong intuitive feelings about what the answers are. And often such intuitive answers will be correct. A large part of independent contracting involves developing an accurate, quick "gut-level" reaction to questions and situations—such as whether or not a company is perceiving reality correctly—that defy detached, objective analysis. Don't be afraid of your subjective, intuitive feelings; put them to use.

At this stage, you are not making any sort of value judgments about the prospect. You are simply trying to determine what the organization's traits, values, and perceptions are, not whether they are "good" or "bad" in any sense. Later on, you will make such judgments to help you decide how, or even whether, you can help the prospect. For now, however, you're just trying to understand as best you can what is really going on with the prospect and why it acts as it does. This sort of information is important in the next step of proposal preparation, because any successful proposal must take into account how a company views "reality" as well as objective reality.

Stating The Problem

Everything you will ever do for a client organization is nothing more than solving a problem. If you're an accountant, you solve the problem of keeping track of where the money is going. If you're an artist, you solve the problem that something needs to be drawn. If you're a financial analyst, you solve problems relating how best to use funds. This means you have to stop thinking that you are going to "provide a service" and instead start thinking in terms of what problems you'll be solving for a company.

Before you can solve any problem for a company, you must be able to state clearly exactly what the problem is, as in "Mr. and Mrs. Jones want to start and run their own bed and breakfast inn, but they have no clue how to do so or where to start," or "The XYZ Company needs several illustrations done for their upcoming brochure." You might be lucky enough to be in a field where problems are that simple and clear, but your skills might be in areas where most problems can't be stated so succinctly. Regardless, it is vital that you develop the clearest, simplest statement of the problem or need you possibly can before attempting to do or solve anything. Your definition of what the problem is determines the goals set forth in the proposal and what actions you propose to take.

In many cases it will be impossible to know exactly what the problem is until you're well into the project, especially if you're proposing to "find out what's wrong" in a company. But you still have to develop the best statement of the problem so you can formulate the rest of your proposal. You can always consider your problem definition at this point as being tentative and change it as you learn more about what's going on in the client company. But without at least tentatively stating the problem, you'll have a difficult time developing a logical plan of action or setting realistic, meaningful goals in your proposal.

One pitfall to avoid is confusing the symptoms of a problem with the problem itself. During the initial interview, many prospects will tell you what their "problem" is, but often they're really telling you only the symptoms of that problem. A prospect might say something like, "Our sales are down 20% from their peak two years ago and we want you to revamp our advertising and promotion efforts so sales can get back to their former levels." This sounds good—the prospect is telling you the problem is with their advertising and promotion—and some independent contractors will accept the prospect's diagnosis at face value. But there could be multiple reasons why sales are down 20% that have nothing whatsoever to do with their sales and promotion efforts. The decreased sales could actually be due to increased competition, reduced spending by their customers, inferior new products, or just lousy management. The only realistic statement of the problem you can make is that their sales are down 20%. By stating the problem this way, you can then develop a plan of action to determine what the cause of the problem is and how to solve it.

The last paragraph doesn't mean you should automatically dismiss the prospect's statement of the problem, just that you shouldn't accept it without verifying it for yourself. It might well be that poor advertising and promotion are behind the 20% drop in sales. Even if they are not, the fact that the prospect would volunteer such a conclusion is an important clue for you. For one thing, it tells you that the prospect is clearly unhappy with their advertising and promotion efforts. That unhappiness might well be the real problem you should attempt to solve rather than trying to reverse their sales decline.

So how do you develop even a tentative statement of the problem based upon your initial interview and preliminary research? Here are some tips based on our experience and those of other independent contractors:

✦ **Your first judgments are often correct.** We've already discussed how your first impressions about an organization's traits are frequently accurate. The same is true for your first impressions about what the company's problems might be based upon your initial meeting with the prospect. While you'll need to do some additional research and thinking to confirm (or disprove) your first judgments, you'll be surprised at how often the organization's problems are obvious from the very beginning to outsiders such as yourself.

✦ **Similar companies have similar problems.** If you've had experience as an employee or independent contractor with similar companies, then you've probably encountered the same type of problems that the prospect is experiencing. Drawing on your experience, you can estimate which problems are most likely, how to identify those problems, and finally how to solve them.

✦ **Most problems—even ostensibly technical ones—are people problems.** In our experience, this is increasingly so as the size of the organization increases and usually involves a lack of communication or rivalries between individuals and groups. If there are problems in an activity that requires communication and coordination between different parts of the organization, some sort of "people problem" is highly likely. Another common "people problem" is the individual that refuses to perform a certain act or make a decision. If there seems to be a bottleneck somewhere in a process or organization, then be alert for such a person somewhere along the line.

✦ **Notice what's missing.** Something that logically *should* be at a company but is not deserves further investigation. For example, the absence of a customer service function, inventory monitoring system, or a quality assurance program may result in any number of problems. A company that is constantly running into cash flow problems might lack an adequate cash flow projection process or a way to quickly alert management when actual cash flow deviates from projections. Missing things can offer big clues about what's going on inside an organization. For example, we remember one company where no one had any personal items—such as family photos—on their desks. It turned out the company forbade such items in employee offices. What would that tell you about the work environment of that company?

✦ **History tends to repeat itself in organizations.** It's no mistake that some companies seem always to be announcing a reorganization or acquisition of other companies. Corporations are legally "persons," and like people have behavioral tendencies and habits. Looking at what a company has done—right and wrong—in the past is often a good way to understand what it is doing now.

> ***"Follow your instincts when dealing with people. Every time I've had a 'bad vibe' with a particular prospect, they've turned out to be a troublesome customer. If I get a feeling that the customer is going to be a penny-pinching type or seems particularly argumentative, I think twice before getting involved in that project."***
>
> **Brian McMurdo, graphics artist, Valley Center, California**

✦ **Don't be afraid of your intuition.** Intuition is disdained by formal management types and business administration professors since it is the antithesis of rational, quantitative business analysis. However, intuition is very real and can be of value in certain situations. Many psychologists describe intuition as a subconscious reasoning process based upon subtle stimuli (tone of voice, body language, whether eye contact is made or not, etc.) that we may not notice on a conscious level. However, the subconscious is noticing such things and forming opinions based upon them. The sort of "aha!" insight we all have from time to time is the result of that subconscious opinion reaching the level of our consciousness. Intuition is not something to base your entire judgment on, but don't entirely discount something you "feel" but cannot fully verbalize.

Setting Objectives

Once you've decided what the problem is, you next need to think in terms of how you would measure a successful solution to it. The usual way this is done is to set objectives or goals for your efforts.

This is the part of your proposal that prospects will be most interested in, since it tells them what they can expect for their money. You'll have to think carefully about what you promise a prospect, since you will be expected to deliver exactly that. The objectives you set will also determine whether a prospect thinks your fee is reasonable or realistic.

A quick way to get in big trouble is to propose ambiguous, hard-to-measure objectives. Let's suppose that you've determined that the problem a company faces is a lack of visibility in the industry and among potential customers, possibly resulting in lost sales for its products. An objective like "establish XYZ Corporation as a recognized industry leader through an aggressive, focused publicity campaign" sounds terrific, but it is also meaningless and an open invitation to conflict with a prospect. Their definition of being a "recognized industry leader" will likely differ from yours, and what is "aggressive, focused" to you might be slack and desultory to the prospect. Moreover, savvy prospects will look at that objective and wonder exactly what you are planning to do for your fee.

It's much better to quantify objectives with measures that you and the prospect can independently verify. The objectives for a publicity campaign for XYZ Corporation could be spelled out in a proposal like this:

- Mail out monthly press releases to the 48 trade magazines covering XYZ Corporation's main areas of business.
- Make monthly telephone calls to editors of those trade magazines suggesting article ideas involving people and/or products from XYZ Corporation.
- Have XYZ Corporation people and/or products as the subject of at least three articles appearing in the 48 trade magazines within the next twelve months.
- Secure at least one radio or television interview in the next twelve months with the president of XYZ Corporation.
- Plan, direct, and execute XYZ Corporation's exhibit at the WidgetCon trade show in October.

Each of these objectives is directly related to increasing the company's recognition in an industry. It is easily possible for you and the prospect to determine if these objectives have been met.

All also have target dates for their completion. The prospect knows what to expect in return for your fee, and you also know what you have to do and by which date. Moreover, listing objectives like this helps the prospect decide what is or is not really important to them. For example, your prospect may not feel that a radio or TV interview with their company president is that important and may ask you to delete that objective in return for a reduced fee.

Every proposal you submit, no matter how simple the task, needs to have at least one clear, quantified objective. Something like "prepare five finished technical illustrations from your hand-drawn originals within a week" is fine if you're a graphic artist handling a straightforward task. However, the more detailed you can make your objectives the better. There's less possibility of misunderstanding that way, and the prospect is more likely to feel they are going to get their money's worth.

Developing A Strategy And Work Plan

The general approach you will take to solve the prospect's problems forms your *strategy*. Using your strategy as a basis, you then plan the exact steps you will take to solve the problem. Those steps constitute your *work plan*.

A strategy is deliberately broad and nondetailed. Going back to our example of XYZ Corporation in the previous section, one way to state the strategy hinted at in the objectives would be "increase XYZ Corporation's visibility in the industry by seeking coverage in the trade press and by participation at trade shows." By contrast, the work plan is detailed. The work plan specifies what will be done, by whom, and by what date. The work plan spells out the responsibilities of both the client and the contractor. For example, someone at XYZ Corporation may have to approve participation at a trade show and select the type of exhibit used. The work plan would specify who at XYZ Corporation would be responsible for that decision and when it would have to be made. As with objectives, the work plan items should be quantifiable whenever possible and capable of being verified by both the client and independent contractor.

In some cases, you might want to include a separate schedule in your proposal even though your work plan will give the dates for tasks to be completed. If there are several tasks and dates involved, it's often better to include a separate schedule clients can consult at a glance instead of forcing them to wade through a lengthy work plan to determine when a task is scheduled for completion. Some prospects may want a detailed chart with milestones indicated or a critical-path diagram for complex assignments.

Develop a schedule you can live with and deliver on. You need to include some extra time for unexpected developments, personal emergencies, and other situations beyond your control. (Few clients will complain if you complete an assignment early!) In your work plan, clearly indicate which items are contingent upon the prospect taking a certain action (reviewing your work, for example) and state the adverse consequences on the schedule of the prospect's failure to perform those actions. You also need to resist pressure (whether overt or implied) from the prospect to agree to a schedule that's overly ambitious. If a prospect says they need a project completed by a certain date and you don't think that's possible, you must honestly voice your doubts. It's better to lose an "impossible" assignment than it is to try futilely to make it work.

The work plan is crucial because it tells the prospect exactly how you plan to solve the problem and what you expect out of the prospect. It shows prospects that you have thought through the problem and have developed a coherent method for solving the problem which they can evaluate.

Budgeting

Some prospects will look at this section of your proposal first. This is where you can justify your fee by showing the value of the different tasks you plan to perform to solve the prospect's problem. While you need to account for every cost you will incur and the amount of profit you want to make on the assignment, there's no need to provide the prospect with that information. Nor does the prospect need item-by-item costs in your proposal; overly detailed budgets in a proposal are often an invitation for the prospect to quibble ("What do you mean, $500 for telephone calls?!?!"). The exact amount of detail you provide depends upon how complex the assignment is and how much detail the prospect wants. For a comparatively simple task (like providing artwork or writing the text of a speech), you may not have to provide any detail at all. For other assignments—such as setting up and managing a small company's accounting function—you might need to assign a separate cost item for dozens of different elements of the assignment.

Most fees will be quoted in two parts. The "fixed" portion of the fee is the amount you will be paid for successful completion of an assignment. The "expense" portion includes things like telephone calls, postage, printing and copying, parts, and other items whose exact cost is impossible to precisely fix in advance. It's a good idea to include your estimate of how much these expenses will be so the client won't feel like you'll take advantage of expenses to produce some extra profit. Another approach is to set a dollar cap on the amount of expenses.

Some budgets might be "provisional" or "contingency based," meaning the prospect will determine what the final cost of the assignment will be based upon their actions (or inactions). Your proposal will probably specify a certain length, time frame, quantity, or other parameters for the assignment. If you get the assignment and the client extends those parameters through their actions (or inactions), then your fee goes up. For example, the proposal may call for the client to provide you with something (such as a working version of the software they are developing) by a certain date. If the client does not do so and you have to alter your schedule as a result, then your fee increases. By the same token, your proposal may budget for three meetings with the client regarding a problem. If additional meetings are requested by the client, then you charge an extra "per meeting" fee. Your budget might include a fixed number of hours or days you will work on an assignment, with additional fees if the client's actions force you to spend more time on it. The key point is that you must not be the one to pay for the client's failures or inefficiencies—the client must. Don't put yourself in a position where you wind up being penalized because of the client's actions. If the client's inaction or mistakes causes you to spend more time and effort on the assignment than outlined in your proposal, then you must be compensated. Otherwise, you're just working extra for free.

If the client objects to such arrangements, use the analogy of a doctor. If a patient refuses to take medicine prescribed by a doctor and needs additional medical treatment as a result, the doctor does not treat the patient for free. If your client refuses to follow your "prescription," then the client will have to pay for the additional "treatment" you must give.

Depending on the project, it may be a good idea to base the total payment due you on a "per item" schedule. For example, you may want to charge for a public relations campaign by the letter or telephone call made on the client's behalf. You may charge by the hour for training or class sessions. As with expenses, include your estimate of how many items you'll charge for, so the prospect has some rough idea of what the total package will cost.

The budget section of your proposal is a good place to include payment terms. It's always a good idea to stagger payments throughout a project based upon the work completed. For example, you might request 20% of the total fee upon acceptance of the proposal and the remaining 80% spread over the life of the assignment as various tasks are completed. A good rule of thumb is to try to get about 20% to 35% of your total fee upon proposal acceptance and have no more than the same percentage remaining to be paid upon completion of the assignment. Depending on the company, you may have to submit invoices for payment, and payment within 30 days or less of submission of the invoice is standard. If the company is new or otherwise seems financially weak, don't be shy about asking for as much as 50% upon proposal acceptance.

Getting The Prospect To State A Preference

You will likely develop several different approaches to solving a prospect's problem as you work on the proposal, or you may have several possible options for the client (for example, a thorough but expensive solution versus a limited but less expensive solution). All of them look equally valid to you, but the important question is which one the prospect would prefer. There's only one way to find this out: telephone the prospect and let them tell you what they prefer. Finding this out before submitting your proposal can tremendously increase the chances your proposal will be accepted.

To use this technique, narrow the possible options or approaches down to no more than two or three. Call the prospect to either set up or reconfirm the day and time at which you will present the proposal in person. Ask some sensible questions, such as who else will be present at the presentation meeting and whether you should bring extra copies of the proposal or other materials. When you have the prospect in a question-answering mood, you can then ask them to indicate a preference. Suppose you offer services as a meeting or convention coordinator and you're proposing to organize a prospect's annual corporate sales meeting. You might say something like this to the client: "I have most elements of my proposal to you ready now, but there is one area where I'm not sure what is best for you. I've selected several possible host hotels, but most of the moderately priced ones lack complimentary airport transportation and have poor, if any, exercise facilities. The typical taxi fare from the airport to most of the less expensive hotels is $25 one way, and there are usually no health clubs or other exercise facilities nearby. The more expensive hotels cost about $20 to $35 more per day but have airport limos and good on-premises health clubs. How important are these two items to your people?" The odds are very high that the prospect will gladly choose one of those alternatives or suggest a third one that's more in line with their preferences. Some prospects will even ask you to make a recommendation for them, especially if there is no large cost differential between alternatives.

Depending on the prospect's mood, your rapport with him or her, and the complexity of your proposal, you may be able to run every significant point—even your fee—in your proposal by the prospect for comment prior to the presentation meeting. In effect, you can get your proposal "pre-approved"! There have been times in our contracting experience when we've had a prospect explicitly tell us to include or delete certain elements of our proposal to insure its acceptance. We've also done the same thing with independent contractors we've used at HighText.

No Surprises!

After your initial meeting, the prospect will have certain expectations about the proposal you will deliver. These expectations will include such things as what you will propose to do to solve the problem, how long it will take, how much it will cost, and even how long your proposal will be. Nobody likes rude surprises, especially prospects who receive a proposal contrary to what they were expecting.

If the final proposal looks like it will differ significantly in some important aspect from what was discussed in the initial meeting (or what you led the prospect to expect), telephone the prospect and discuss the situation candidly. A good way to start the conversation is to say that the proposal is shaping up differently from what you anticipated, and you'd like to make sure you did not misunderstand or misinterpret the information you got from the initial meeting. Be prepared to explain the factors and reasoning behind your position, but also be receptive to new information from the prospect and willing to revise your proposal based on the prospect's responses.

However, if you still feel strongly about a point, such as the length of time it will take to do an assignment or the best method of doing it, politely but firmly stick to your guns. Don't let your desire for an assignment cause you to offer a proposal that you can't live with or deliver on. You'll be stuck with the consequences (i.e., the blame) if you can't perform what you agree to in your proposal, and this can involve a lot of headaches, actually losing money on the assignment, or even getting sued by an angry client. If the prospect isn't willing to consider a proposal reflecting changes you feel strongly about, it's better to decline to submit any proposal. Both you and the prospect will be spared a lot of grief.

Writing The Proposal

The prospect's characteristics and the complexity of the task you propose to do determine the form, length, tone, and amount of detail in your proposal. A simple task for an informal organization can be proposed in a friendly letter. By contrast, a complex series of tasks to be performed over a period of time for a large, bureaucratic organization will often require a detailed, formal proposal of several pages.

There are certain elements common to any proposal regardless of length. Not all of the items below belong in every proposal (having an introductory section in a letter proposal is clearly ridiculous, for example), and your proposal doesn't have to follow this sequence. As you become more experienced as an independent contractor, you'll develop your own proposal style and know how elaborate a proposal is appropriate for various projects. However, this is a good checklist for developing your first proposals:

✦ **Introduction or Opening.** This is appropriate if you have a lengthy or complex proposal. Don't try to do any selling or convincing here; instead, describe how the proposal is organized and perhaps briefly summarize your conclusions (don't mention your fee, however). If you include this section, keep it as brief as possible.

✦ **Background Summary.** This is usually necessary only if you're proposing a long or complex assignment. This is to demonstrate your understanding of the prospect's situation and the basic assumptions you will be operating under during the assignment.

✦ **Statement of the Problem.** This is where you state the definition of the problem(s) that you have developed. This section may require considerable diplomacy and tact on your part if the organization's management or policies are the problem! The best course is to avoid placing blame for the past and instead focus on new opportunities to correct the situation. Again, if your task is simple and well-defined you can omit this section.

✦ **Objectives.** Here is where you put the list of what you will accomplish for your fee. Even a simple letter proposal needs to clearly state at least one objective.

✦ **Strategy and Methods.** If the assignment is complex or lengthy, tell the prospect what strategy you will follow to perform it. Also, briefly describe your working methodology (interviews with the prospect's employees, statistical analysis, object oriented programming, etc.) and why such methodology is the best approach for solving the problem.

✦ **Work Plan.** This is the work plan you've previously developed. Be sure to include the responsibilities the prospect will have in regard to the assignment, such as making key people available for meetings with you or providing information or reviews by certain dates. It's a good idea here to ask the prospect to designate one person in the organization to be a "contact person" through which you will deal with the organization and the organization will deal with you.

✦ **Schedule**. While your work plan will include target dates and deadlines, it's often helpful to summarize them in a separate section (especially very important ones). For more complex assignments, some sort of time line or milestone chart might be appropriate.

✦ **Budget.** Remember our earlier warning about getting into too much detail about the budget? It's best to assign a complete cost figure, including a percentage of total overhead, to each of the main tasks you propose to do. For example, HighText typically breaks technical manual preparation assignments into writing, editing, design, page layout, and technical illustration sections in proposals and places a separate cost on each section. We also include contingency terms for such things as additional meetings with a client and changes made to previously approved materials. This is also the place to specify payment terms.

✦ **Guarantee.** You wouldn't buy anything from a stranger that didn't have some sort of warranty or guarantee, would you? You don't need to put in an elaborate guarantee loaded with legalese, but something like "I promise to do all work to XYZ Corporation's satisfaction or else XYZ Corporation owes nothing" will go a long way toward getting your proposals accepted, especially if you are a new contractor without satisfied clients you can use as references.

✦ **Acceptance Language.** This is an essential item for any proposal. Since your proposal is intended to be a contract for an assignment, include a sentence or two allowing the prospect to agree to your terms. Again, there's generally no need for formal language here. A phrase like "I'd welcome the chance to perform this assignment for you. If the terms of this proposal are acceptable to you, just sign in the space below to indicate your agreement with them and I'll start putting the plan I've described into action. Thank you for considering this proposal." Immediately below this would be a line for the prospect's signature and the date. While some lawyers might want more inclusive or formal language used, a signature below such phrases would indeed constitute acceptance of the proposal and agreement with its terms, and the proposal would be converted into a binding contract. Since the prospect will want a copy of your proposal for their records, submit two copies of the proposal. The prospect signs both copies, returning one to you and retaining one.

If you're uncertain of how long or detailed a proposal should be, we suggest erring on the side of brevity and conciseness. If the prospect wants more details, you'll be asked to provide them; if you provide a lot of details the prospect doesn't want, your proposal might not get completely read. The objectives, work plan, and schedule sections should always be fairly complete, but details can usually be omitted in the other sections without too much trouble. You probably will have a good idea of how complete a proposal the prospect wants based on your meeting and telephone conversations. If there is some doubt, it's best to call the prospect and ask if they want a lengthy, detailed proposal.

While there's no need to make most proposals "fancy," with elaborate layouts or typefaces, it is essential to make sure the proposal contains no grammatical errors, misspellings (especially of names!), or poorly written, confusing text. The layout of the text on the page should be neat and professional, with ample page margins and "white space" between sections (the prospect may want to make notes or comments on the proposal, especially if more than one person is to review it). If your proposal looks sloppy, then the prospect may well conclude that your work would be equally sloppy.

Presenting Your Proposal

It's usually best not to submit your proposal to the prospect by mail or fax. If possible, give it directly to the client in a *presentation meeting*. Depending on the client and the task you propose, the presentation meeting might be a simple one-on-one meeting taking only a few minutes or a lengthy session with several people who ask pointed questions about the proposal. Even if the prospect indicates it's okay to submit your proposal by mail, always ask for a meeting to present the proposal. The client will probably have some questions about what you're proposing—or even some outright objections—and it's easier and more effective to handle these in a meeting.

Let the prospect decide how long the presentation meeting should be. When you call the prospect to arrange the presentation meeting, ask how much time you should take. *Then stick to that schedule.* Remain flexible should the prospect want to extend the meeting, but don't do it unilaterally. Your written proposal should be able to stand on its own. If a point is crucial, it should be in the written proposal instead of your verbal presentation.

The exact format for your presentation will depend upon the nature of your proposal and the prospect. The most common format is to open with a quick summary of the main points of your proposal, answer questions from the prospect, and close with—and this is very important!—a call for the prospect to accept the proposal. More important than the format is your attitude. You *must* project enthusiasm and competence. The presentation is a final opportunity to sell—or "unsell"—yourself. If you look uncomfortable or uninterested as you discuss the prospect's problems and your plan to solve them, you'll never get a chance to put your plan into action.

Prepare for your presentation by making a mental list of the highlights of your proposal. These highlights will usually be your statement of the problem, the objectives you've set, the methods you'll use to achieve those objectives, the schedule, and the budget. Don't get bogged down in detail here or read directly from your proposal; instead, your purpose is to give the prospect enough of an overview of your proposal to initiate discussion about it with you. The shorter you can make your summary, the better. A good rule of thumb is that no opening summary should take more than five minutes unless you're discussing an extraordinarily difficult or complex assignment. Try whenever possible to get your summary down to less than two minutes.

After you give the summary, you will usually get some questions from the prospect. Most questions will be "informational," seeking information about what you propose to do. Often the answers to such questions will be in your proposal. In such cases, you can say something like "That's a good question, and the answer is on page xxx of the proposal. Briefly, . . ." and then answer the question. It's also a good idea to repeat or slightly rephrase most questions you're asked. This gives you the chance to better compose your answer, and you appear more thoughtful when you don't immediately blurt out an answer.

A lot of the informational questions you will be asked can be anticipated. Look at your proposal as objectively as you can and zero in on those areas—the problem definition, objectives, methods, schedule, and budget—that are likely to be of most interest to the prospect. You also need to look for the weakest links in your proposal and be prepared to justify your reasoning and assumptions. If you do get a question that you can't answer, don't try to fake a response. Admit that you don't know the answer and offer to find out and get back in touch with the questioner. Your candor and willingness to locate an answer will score more points for you than an improvised response that the prospect might realize is incorrect.

Two types of informational questions you'll need to finesse involve your work methods and fee. Just as in the initial meeting, you don't want to reveal too much about exactly how you will solve a problem, since you might wind up negating the need for your services! It's best to keep your responses to such questions general instead of specific. If repeatedly pressed for "how to do it" instructions, don't be afraid to respond with "My exact methods and techniques are my trade secrets and proprietary information. They are essential to

my business, just as your company has information it must keep confidential, so I can't answer as completely as you might like. However, in general terms . . ." If the prospect is seriously interested in using your services, they won't be offended by such a response. If they are, then they probably were just fishing for information and weren't serious about using your services anyway.

Questions about your fee should likewise be deflected, since how you determine your price or how much profit you make is really none of the prospect's business. You can reply by saying your rates are competitive with people of similar abilities or experience, that you feel your fee is commensurate with the scope of the work to be performed, or that your fee is equitable considering the value of the work you are to perform.

You will run into some hostile or aggressive questions from time to time, particularly when you're giving your presentation to more than one person. In such situations, hostile questions might be little more than posturing among members of the group. In some cases, you may have touched a nerve with a hostile questioner—for example, your proposed solution might be in direct opposition to a course of action advocated by the questioner. Or you may be dealing with somebody whose natural manner of asking any question is brusque and curt.

Whatever the reason, don't succumb to the temptation to respond in an equally hostile or aggressive style. After all, such people are doing you a favor; they would surely raise their objections to your proposal sometime, and by asking hostile questions in your presence they're giving you the chance to respond to their objections. One effective technique for handling a hostile question is to say something like "That's an interesting question. To help me give you a useful answer, could you tell me some of your concerns that prompted it?" Don't challenge or try to rebut the questioner, but do try to find out what is really motivating the question. For example, the questioner might say they tried a similar solution in the past that did not work. You can reply by showing why your solution will work (you may have to ask more questions about the previous solution before you can tell why your solution will work). On the other hand, if you get a fuzzy, indefinite response ("I really don't know what I don't like about it . . . I just don't like it, that's all") then you can safely assume that the questioner is either confused or just grandstanding. In such cases, express empathy ("I know my proposed solution can seem complex, yet . . .") and then restate the main reasons underlying your recommended course of action. If you've put in the effort necessary to prepare a good proposal, you should have plenty of facts and figures at your disposal to adequately counter any objections.

You'll eventually get a hostile question from some prospect involving your fee. The question may be phrased as a statement such as "We can probably find someone to do this cheaper, you know." When this happens, aggressively defend your price by pointing out that you're offering a unique package of ability and experience: "I understand that cost is an important consideration in your decision, but I don't think it's the only criterion you'll use. I know you could find someone willing to do this assignment more cheaply, but I think you'll have a hard time finding someone able to do it as well or better for the fee I'd charge you. I honestly feel you'll get full value for every cent you pay me." (We'll discuss fee negotiation in the next section.)

What about situations where you fundamentally and irreconcilably disagree with a questioner? You won't get anywhere by arguing, but that doesn't mean you should express false agreement either. Politely but firmly stick to your guns by saying something like, "I respect your opinion on that, but I have to disagree with you. My experience has been that . . ." and then restate your position. Odds are the prospect will respect your honesty and conviction more than they will be upset over any disagreement.

Finally, ask the prospect for a decision at the end of your presentation by saying something like, "I'd like to begin work on this assignment as soon as possible. Your signature at the end of my proposal is all I need to get started." In sales, this is known as *closing the deal*, and it's equally essential for successful independent contracting. You have to directly and unambiguously ask for the assignment. If the prospect is unable to make a commitment at the end of your presentation, ask for a date when they will be able to make such a commitment. One way to spur the prospect to take some action is to include a time limit, such as 30 days, in which the prospect can accept your proposal. Make it clear, in writing and verbally, that you may have to revise the terms and conditions of the proposal—including your fee—if the prospect wants to accept it after that period.

Follow up the presentation with a written thank-you note and check back near the date when the prospect said they would be able to make a decision on your proposal. There will be a certain percentage of cases in which you will never be able to get a firm decision out of a prospect. If a couple of follow-up calls to a prospect fail to produce a decision, it's best to devote your energies to other prospects and proposals. Some people you encounter will be unable to make any sort of decision, while others are uncomfortable in saying no to someone. There's little you can do to positively influence a prospect's decision after your presentation, and if a favorable decision is not forthcoming by the prospect's original deadline it usually won't be coming at all. However, do save your proposals and supporting materials. Sometimes a prospect will contact you months after your original proposal and ask if you are still interested in an assignment. (This often means the prospect tried another contractor that didn't work out.) You might want to re-approach the prospect a year or so later; there might be new employees more favorably inclined toward your proposal. Finally, parts of previous proposals can sometimes be re-used in proposals for other prospects.

Negotiating During The Presentation

During your presentation, you'll probably get some questions like, "Is it possible to have this completed earlier?" or "Could you also handle the printing of the brochure for us?" Such questions are actually invitations to negotiate the terms and conditions of your proposal. This is a common situation, and for many independent contractors this is a normal part of their presentation.

Keep in mind that successful negotiations are based on a *quid pro quo* between all parties involved. If the prospect wants something additional from you, you should get something extra from the prospect. By the same token, you must be willing to make concessions if you want concessions from the prospect. And it's important that you and the prospect feel good about the resulting agreement. If either of you (or both!) feel like a loser after the negotiation process, there may be some leftover resentment and ill will that can mar the assignment. An agreement that seems reasonable and fair to both parties—that is, an agreement where what you give up is equal to what you receive back—is the easiest kind to implement and perform.

Be ready to take the initiative when a prospect wants to negotiate. One good way to do so is to build some "slack" into your proposal so you can quickly make what seem to be unilateral concessions. For example, include more time in your schedule than you think the project will actually take or budget for more copies of a final report than you promise in your proposal. This lets you appear to be eager to please and helps build good will for you. Your willingness to compromise may encourage the prospect to do the same. (And if you don't have to make any concessions, then you're still ahead of the game!) Once you've made a couple of "free" concessions, however, it's time to dig in your heels until the prospect shows a willingness to meet you halfway. If you agree to an accelerated schedule and to provide additional copies of the final report, wait until the prospect offers a concession before making additional concessions on your part.

One place where you should always hold the line is your price. Here is a rule that has served us well in our contracting assignments:

Never reduce the fee quoted in your proposal unless you also reduce the services you provide.

There are a number of reasons why you should always follow this rule. Perhaps the most important is that a willingness to reduce your fee without reducing the services you perform demonstrates that you didn't really believe your services were worth the fee you proposed. A fee reduction can actually reduce your attractiveness to the prospect—if you don't feel you're really worth the amount in your proposal, how can the prospect be sure you're worth even the reduced fee? In all future proposals, the prospect will expect you to reduce your fee and never take the amount quoted in your proposal seriously. Moreover, many prospects will interpret your willingness to concede a fee reduction as a sign of major weakness on your part (such as being desperate for the assignment) and will drive harder for concessions on other proposal items.

If the prospect indicates your fee is too high, politely but firmly say it is justified by the services you are proposing. But then tell the prospect the fee can be reduced if a lower level of services is acceptable. In our technical manual work at HighText, we have reduced our quoted fee if the prospect could accept shorter manuals ("shorter" meaning fewer words, not just fewer pages) with fewer illustrations and less in-depth technical material. Your work plan is the place to start in determining what the prospect can live without. You will know what items are essential to completing the task and which are just "nice" to have, and this will let you make recommendations on what can be eliminated to the prospect. Offer to come up with an alternate work plan with a lower fee, if necessary, but don't reduce your quoted fee without a proportional reduction in the work you will actually do. Your services are not some fungible commodity like wheat or iron ore; instead, they are unique. No one will bring the same mix of education, intelligence, experience, and skill to the assignment, and there is no direct substitute for you. (We'll have more to say about the whole issue of pricing later, but for now just remember that the biggest mistake most independent contractors make is *underpricing* their services.)

If the changes you and the prospect agree upon are minor, you can make changes directly on the copy you keep and the one the prospect keeps. Both you and the prospect should initial all changes on both copies. If the changes are extensive or complex, it's best to prepare a new proposal incorporating those changes.

After reading this chapter, preparing and presenting a proposal might seem like as much work as the actual project. Your first few proposals will indeed probably take a lot of time and effort. With practice, however, you will develop your own system of preparing proposals for various assignments. Experience will tell you how lengthy and detailed a proposal should be, and you'll be able to analyze a prospect's situation more rapidly and write proposals in less time.

Chapter 6
Making It Work

Getting a signature on a proposal you submit is a cause for celebration on your part, since you've converted a prospect into a paying client. In many cases, the hardest part of your job may be done at this point! Performing the actual services called for under your agreement with the client can be easier than locating the client and "selling" your proposal to them.

However, there are pitfalls you should look out for, especially if you're a new independent contractor. The attitudes and habits that make you a good permanent employee can actually work against you when you're a contractor (and vice-versa, for that matter). You have to correctly understand the exact nature of your relationship with the organization that uses your services on a contract basis, and you also have to make sure the organization correctly understands your role and standing. In this chapter, we'll look at how to conduct yourself as an independent contractor and make your relationship with the organization work. And, as an example of what can go wrong and how to handle it, we'll dig one of the bigger skeletons out of the High-Text closet and examine it at the end of this chapter.

You're INDEPENDENT And A CONTRACTOR

This is a good place to once again stress that you're an *independent contractor*, not an employee. We can't place enough emphasis on those words "independent" and "contractor," because you can't succeed as an independent contractor unless you grasp the full ramifications of those two words.

Since you're independent, you don't need anyone else to provide you with the motivation, skills, or resources necessary to do a good job. You are able to assess your performance, know when you are not performing satisfactorily, and are able to take steps to improve your performance. You take the initiative in performing tasks, and understand that a good effort or a good excuse isn't an acceptable substitute for desired results. You're comfortable shouldering the responsibility for completing an assignment satisfactorily; this is good because that responsibility is inherently yours.

Since you're a contractor, your relationship with the organization is governed by a written agreement and is one between equals. The company using your services cannot unilaterally change the nature of the assignment you'll be performing, the schedule for its completion, or how much you'll be paid. By the same token, you can't unilaterally extend the schedule for a project or perform different services than you agreed to do. If you do, you may get more than just "fired"; you might wind up getting sued for breach of contract.

We've been hammering these points home in previous chapters, and for good reason. Getting yourself out of an employee mindset and into thinking and acting like an independent contractor is extraordinarily difficult for many people, especially those who have spent years as an employee. These people are skilled, experienced, and would seem to have all the resources necessary to become successful independent contractors. But they fail because they persist in viewing the organization as "their company" and their contact person in the organization as "the boss." These people wait for the organization to initiate activities and direct the assignment instead of doing it themselves. They are passive, deferential, and wait to be told how they should proceed. They believe the organization is interested in them as individuals instead of for what they can do; they assume the organization will understand if they are behind schedule because of something unexpected or personal matters.

They think the people they deal with at the organizations are their friends just because those people act in a friendly manner toward them.

Such "independent contractors" wind up frustrating most of the organizations they are supposed to help. Your independence—your presumed ability to work without close supervision, motivate yourself, and take responsibility for outcomes—is a big reason companies use independent contractors instead of hiring a permanent employee. Clients want to review your work, not supervise it. If you expect a client to tell you how something should be done, when you should do it, and whether you doing a good job or not, you're acting like an employee instead of an independent contractor. Moreover, no matter how well you get along with the people in the organization, they and their employer are not your friends and won't "understand" like a friend would if you fail to do what you promised.

You're an independent contractor, not an employee. Don't forget it!

Keeping Records

Since your relationship with an organization is contractual, you need to keep written records of everything affecting your contract. Suppose the client wants to extend the completion date for a project by a month. If that's agreeable to you, you can indicate agreement by dropping the client a short letter saying that you agree with their request to extend the completion date. *Keep a copy for your records.*

Similarly, any change to the terms and conditions of the accepted proposal should be put in writing with one copy to the client and one for you. It also helps to summarize in writing the results of any meetings, tests, or discussions you may have with the client. You don't do this because you distrust the client, but because people forget, are transferred to another position, or are overwhelmed by their workload. Having things in writing makes it easier to know exactly what is to be done, who said what when, and lets new people dealing with the project rapidly learn what's going on. Putting things in a letters lets clients know you understand what is to be done and also lets clients change their minds or correct errors.

Make notes of any telephone calls or conversations you have with a client regarding major issues affecting your contract or the project. This doesn't mean to write down every minute or trivial thing said, but that all important points should be noted in writing and then sent in a letter to the client. If there is something that you feel should be kept confidential (like a subordinate who expresses unhappiness with his or her boss), you can record it in a "memo for the record" which you do not send to the client. Some independent contractors keep a journal or "professional diary" in which they keep track of such information.

There will be too many details on most projects for you to carry around in your head. Putting things in writing and sharing them with your client will help both of you make the assignment go much smoother.

Asserting Your Independence

You're not the only one who must recognize the difference between being a permanent employee and an independent contractor. Your client must do likewise. The client who insists on treating you like one of the "hired help" is a common problem. Clients who do so are not being malicious; they're just forgetful. You'll need to politely but clearly assert your independence in such situations.

A frequent problem is the client who makes requests that are contrary to the terms and conditions of your contract. If the requests are minor and you can agree to them without any trouble, you should indicate your verbal approval and then send a confirming letter. However, some client requests are so sweeping they materially alter your agreement, such as a request for a much shorter deadline or for you to do something quite different from what you proposed. If you get one of these "blockbuster" requests, say something like, "That's a bigger change than I can simply agree to. What you're asking is going to cost me some money and will make it hard for me to keep my commitments to my other clients. To do this, we'll have to amend our agreement to cover the additional time and effort this will take." Then offer to submit a bid on your fee for altering the contract. You can submit this in a letter form with a place for their signature to indicate acceptance, much like your original proposal.

If the client is reasonable, they will understand why you're taking this approach and handle it in a professional manner. Fortunately, most clients will be reasonable. There will be a few, however, who will be upset at the possibility of having to pay you more because they want you to do more or to do something different from your agreement with them. Reactions we've encountered have run the gamut from a wheedling "I'm not asking all that much" to an angry "Aren't we paying you enough already?" Don't lose your cool just because the client does, though. Just say that you both had reached an agreement, that your fee and schedule were based upon that agreement, and if they want significant changes in that agreement it should be renegotiated, and you're willing to do that. Ask them how they would feel if you had asked them if you could do less for the same fee or get a higher fee for the same work. A few clients will persist in viewing you more as an indentured servant than an independent professional and refuse to even consider your position. In this case, be willing to walk away from the project. It's very unlikely you'll get much accomplished with such people, and your time and energy is better spent elsewhere.

More common are situations where a constant stream of small changes or requests can, over time, wind up changing the nature of the assignment. This is almost never a deliberate strategy by the client, but rather a failure of the client to realize that your working relationship is governed by a written agreement. You obviously will not make an issue over minor breaches of your agreement by the client (just as most clients will not do so over minor breaches on your part). To keep the project moving, you'll look the other way when the client fails to act in a certain way or makes requests not covered in your agreement. By themselves, such breaches and requests will not be significant.

When several happen in a short time, however, you may realize one morning that you're in a very difficult situation.

After having gotten caught in a couple of situations like this ourselves, our policy is now to immediately err on the side of caution when we sense that a project is starting to "get away" from us. One method that's been effective for us to send a friendly letter pointing out what seems to be a trend (slow review of materials submitted to the client, a gradual "stretching" of the schedule, etc.) and expressing hope that this can be corrected before it begins to have an impact on the project. The client will often appreciate having such concerns brought to their attention and will take steps to get things under control on their end. If the problem continues or even worsens, make each subsequent letter more pointed and forceful than the last. If the problems are really bad and you can't get relief, be prepared to withdraw from the project. The last section of this chapter is an example of such a project and how it can be handled. The best policy, however, is to prevent the situation from getting so bad in the first place.

Renegotiating

What do you do when the customer wants to renegotiate the deal? Sometimes the requirements of a particular project may become so different from the original specifications that the entire job may need to be renegotiated. If your client initiates a renegotiation, ask questions to find out why. Does it mean they no longer have funds to complete the work? Or does it mean they found more money and want to do more? If only minor changes to the scope of work are involved, then it may not be a problem. You can reach an agreement on the new work to be done and the price to be paid, and write an addendum to the contract for both parties to sign.

Sometimes a customer wants to cancel a job before it's completed, due to lack of funding or some other reason. This is unfortunate, but as long as the client is willing to pay you for the work completed and you can both agree on a reasonable amount as a final payment, there's not much that you can do about it. This has happened to us at HighText on a couple of software manual projects. In both cases, internal politics at the client companies led to the job being canceled after we had ramped up for a major effort. This can be tough if you're counting on those revenues that never materialize. It's always a good idea to have contingency plans at the ready; never bet your business on a single job!

An even trickier situation arises when you find yourself in trouble on a project and want to renegotiate the payment terms. Perhaps the client has been gradually sneaking in a few extra tasks here and there, and suddenly you find that you're spending a lot of time but not making any money. Or, even worse, maybe you severely underbid the job and are now in financial trouble. Either way, this is a tough call.

If you are in severe trouble, first try to analyze the situation to understand the cause. If the scope of work or other parameters have changed significantly, you may have valid grounds for renegotiating. If you simply underbid the job, you don't really have a leg to stand on, unless you have an especially understanding and forgiving client. Once you have a signed contract, the client is under no obligation to renegotiate unless the scope of work has changed. As we've stressed earlier, you must be very careful with fixed-price jobs—you have to make absolutely sure that your costs are covered. If you can't precisely define the scope of work involved in a project, you may be better off with a "time and materials" contract, in which you are paid for every hour worked and for all the expenses you incur.

What If the Customer's Wrong?

Sometimes the client will want you to do something a certain way . . . and you know it's the wrong way! This situation requires tact and diplomacy on your part. The only thing to do is to explain your position and press your case as best you can. If, after hearing your opinion, they still insist on doing it their way, then do it their way. After all, the client is paying the bill. If you think that there may be a dire outcome that could hurt your professional reputation, then politely withdraw from the job.

One possible way out is to show the client a sample using their idea, and another doing it the way you think it should be done. When presented with a choice in this way, the customer will often choose "correctly." If not, either do it their way or decline the assignment.

Resistance And Hostility

You will eventually find people in a client organization who intensely resent you for some reason you can't explain. These people might be angry because they wanted to do the project you were assigned. They might feel threatened or insecure because of you. (Were you hired because their boss is unhappy with them???). Or they might be jealous of your independence or how much you're getting paid. Whatever the reason, they make sure you know that they don't like you. Others will have passive aggression down to an art form. They are friendly to you, but you need them to complete a task before you can proceed and nothing . . . ever . . . gets . . . done.

As long as their hostility doesn't affect your ability to complete your assignment, the best thing is to just ignore such people. After all, you're not doing the project because you want to be loved; you're doing it so you can get paid. If their hostility and resistance impedes your ability to complete your task, the matter is much more serious since it can affect your income.

Here is where keeping good records can be invaluable. On one book-length technical writing job HighText handled, it became obvious that the person responsible for reviewing the materials we submitted had a real problem. We would submit several chapters and then hear nothing from the client for *months.* Their delays were interfering with the project deadline and also with our payment schedule, since we couldn't complete the work without the client's review.

We had kept a file of every piece of correspondence with the client, so we finally wrote a strongly worded letter to the supervisor of the person we had been dealing with, and attached copies of the correspondence, which clearly showed that we had been responsive and their person hadn't. This got results, and we were able to complete the project. Of course, we sacrificed the relationship with the original contact, but at that point it wasn't worth much to us anymore.

Keeping Track Of Your Finances

We definitely recommend starting out with an accountant who has experience working with independent contractors and small businesses. Have the accountant set up a bookkeeping system for you and *take the time to understand it.* Your financial recordkeeping is the lifeblood of your business!

Since computers are so cheap and easy to use nowadays, it's imperative that you do your finances on a computer. There are many off-the-shelf accounting packages that are tailored to contractors and small businesses. Many incorporate useful features such as invoicing, estimating, payroll, purchase orders, etc.

If you don't keep good, organized financial records your odds of success are slim to none. You need to have enough knowledge about the financial workings of your business to talk intelligently about them with your accountant. If you are resistant at first to learning the principles of accounting and adhering to them, you might be surprised to find yourself actually enjoying this aspect of working on your own. Neither of us had any particular interest or love for accounting when we started out, but we both have developed a healthy respect for it and enjoy cutting invoices and running the monthly reports. Computers have made it relatively easy.

Invoicing

Most of your clients will ask that you submit invoices for any payments due you. An invoice is a simply a written request for payment (or what most of us call a bill), and is required under most accounting systems (especially those of larger clients) so payments to outsiders can be tracked and properly charged against the correct accounts.

An invoice can be remarkably simple. Many computer accounting packages have an invoicing form built in. If you're not using a computerized accounting system, then you can make up your own form. Include your name and address, as well as that of the client, and the date you are submitting the invoice. You should include a "customer reference" on the invoice, which refers to the document under which you are claiming payment. If the client accepted your written proposal, the reference would be something like "Page 4, section 8 of our agreement on March 15, 1993." If the client issued a purchase order for your proposal, the reference would be "your purchase order # AZ458." The purpose of the reference is to let the accounting department determine whether or not payment of your invoice is authorized. It's a good idea to include the name of your contact person at the client company so the accounting department can quickly check with him or her should they have a question about the invoice. Include a written description of the work you are requesting payment for, and indicate when payment should be made. Thirty days is the usual period, and you state this with a phrase such as "Terms: net 30 days."

Each invoice should be uniquely numbered. The simplest system is to number all invoices consecutively. Another approach is to use the date as the invoice number; if you send out more than one invoice on a day, use letters as prefixes or suffixes to distinguish the invoices.

Determine from the client who should receive your invoices; usually this will be either your contact person or the accounts receivable department at the company. Always keep a copy of the invoice for your records. The adjacent figure shows a typical invoice that should serve most independent contractors well.

Invoicing is one of the most important elements of working on your own. If you don't send out invoices, you won't get paid. We're often surprised at how late many of our suppliers are with their invoices. (We love it, because they're granting us an interest-free loan!) Send out your invoices as soon as you can—don't let them pile up.

Invoice #	609
Date:	February 15, 1994
To:	TechTronix, Inc. 138 Great Western Way Santa Clara, CA 95050
From:	HighText Publications P. O. Box 1489 Solana Beach, CA 92075
Contact:	Mark Johnson, director of corporate communications
Reference:	Paragraph 12 of our agreement of October 5, 1993
Services Provided:	Writing and editing of first draft of service manual for DB127 signal analyzer and submission of all preliminary technical illustrations for manual
Amount Due:	$7500.00
Terms:	Net 30 days from date of this invoice

Late Payments

In our experience, if you stay on top of your invoicing and bill collecting—and choose your clients wisely—you'll have very few problems with uncollected bills. Unfortunately, you may sometimes run into a client that pays slowly or not at all. This can have several causes. Sometimes invoices do get lost under a pile of papers on someone's desk. Some companies take longer than 30 days to pay as a matter of course—we've found 45 days to be about average for larger companies. And in some cases the company is short of cash and won't/can't pay your invoice. Whatever the case, you must take quick, firm action whenever an invoice is overdue.

Many companies will take the full 30 days to pay your invoice, and will not put your check in the mail until the thirty-first day! It's good idea to wait until about five days after the end of the 30 days before making a query. You can phone your contact person at the client or the client's accounting department and ask about the invoice; a good place to start is the accounting department. Ask if they have received the invoice for payment, and refer to it by date and number. If they have received it, ask when payment will be made. Don't settle for an indefinite response like "sometime next week." Instead, insist upon a firm date when payment will be made; say that you want to receive payment in no more than ten days. If you can't get such a commitment from the accounting department, call your contact person at the company, explain the situation, and ask them to help.

The accounting department might tell you that the invoice was never submitted for payment. This means that someone dropped the ball and the invoice is probably sitting in someone's "in box." The most likely culprits are either your contact person or his/her boss. Regardless of what explanation you get, say that you want to receive payment in ten days. Be friendly but firm on this point.

Follow up your phone conversations with a letter to your contact person at the company. Express thanks for their assistance in resolving the problem, but point out that your agreement with them calls for payment within 30 days. State that if you have not received your check by the date promised, or within another ten days if no date was promised, that you will stop working on the project until the check is received. Tell the client they will be liable for any additional charges and delays that may result if you have to stop work due to no payment being received. Tell them you would regret having to take these steps, that you're confident the problem will be resolved, and that you look forward to working with them on the remainder of the project.

If a check is only a few days late, the odds are good that a minor, easily correctable problem is the cause, especially if your client is a medium or large firm in strong financial condition. It's easy for someone to forget to process an invoice or for an accounting clerk to make an error in entering the invoice for payment. You might be told that payment beyond 30 days is the standard policy of the organization. Your response can rightfully be that you should have been informed of that when you submitted your proposal calling for payment within 30 days. How strongly you want to push the issue depends on whether you want to do additional work for the client. If you do, and the payments eventually arrive on a regular schedule and do not disrupt your finances, it's best to let the matter pass. You might want to increase your next proposed fee to cover the delays or, if you're currently being paid by the hour, to add an extra hour or two to your next invoice as a sort of "finance charge." After all, a bank would charge the company interest on any loan, and an unpaid invoice is a "loan" to the company.

If the company fails to get your check to you by their deadline or within ten days, the time has come for stronger measures. Contact the client's accounting department and your contact person, tell them you still haven't received your check, and ask them what they are going to do about it. You will probably get multiple excuses and reasons for the delay, but this is not a time to be sympathetic or understanding. Whatever reasons they have for not paying are their problems, not yours—your problem is that you're not getting paid as you're supposed to. Nothing else should matter to you at this point. The client is not doing you a favor by paying you; they're fulfilling a legal obligation and you're well within your rights to demand immediate payment. Don't be concerned about jeopardizing your reputation or future relationship with the client. Your object is to be paid money, not promised money; if a client can't pay you as they agreed to, they're not the sort of client you need.

While screaming and yelling won't get you very far at this point, firmness and decisiveness will. Tell both your contact person and the accounting department that they have breached your agreement with them, that you will do no further work

> **Cash Flow**
>
> ***"In my line of work—book publishing—some jobs take forever to bill out. You have to be prepared to go for four to five months before being paid. Many freelancers just starting out forget to plan for this cash flow bind. You need to have some solid funding when starting out to carry you until the money begins to come in."***
>
> **Lynn Edwards, Bookmark Book Production and Electronic Prepress Services, San Diego, California**

on the project, and that the agreement will have to be renegotiated should they want the project completed. This is also the point to say that you are willing to take legal action to secure payment. Demand your check within five days. Put all of this in writing and send it to both the accounting department and the contact person. Make follow-up calls and letters each five days until your invoice is 60 days past due.

If 60 days pass without payment, you're justified in suspecting the worst about the client and it's time to take off the gloves. Start calling both the accounting department and contact person on a daily basis if you haven't been paid after 60 days. Persistence and increasing the "threat level" are called for at this point. In our experience—as well as those of other independent contractors we know—it is the most aggressive creditors that get paid by cash-strapped companies. If you are willing to be "understanding" and not press for payment, the company has no real motivation to pay you. If you're on the phone daily and make it clear that you're willing to take legal action if necessary, then you give the company an incentive to pay even if it means they have to postpone paying others. Don't feel guilty about this.

What if they still don't respond after 90 days or so? If the client company is local and the amount owed you is relatively small, your local small claims court can be a good way to press any legal action if necessary. Call your local small claims court to determine the dollar limit on cases they will accept. For larger amounts, you will have to retain a lawyer to press action for payment. If the company is not local and the amount owed you is small, then it may make more sense to just charge the amount off as a bad debt instead of filing suit.

As mentioned in earlier chapters, the best way to avoid such difficulties is to be selective about who you take on as a client. New and smaller firms are the biggest payment risks, and you should ask for a healthy portion of your fee—such as a third—in advance before starting *any* work on the project. It's better to lose a project for insisting on a healthy first fee payment than it is to "win" a project for which you wind up not getting paid in a timely fashion or even at all.

Ethical Conflicts

Professional ethics is an important consideration for an independent contractor in any field. How do you react when a client asks you to do something that you consider "over the line" ethically? This could be a blatant request—perhaps the customer asks you to falsify some numbers or lie about something—or it can be a more subtle suggestion to put a better face on something than you feel is justified by the facts. Whatever the case, it's up to you to determine what's really going on and to defend yourself against it.

The best defense is to scrupulously pursue the truth of the matter, and do so to the client's face. For example, if you are

Ethics

"I had a client who wanted me to lie to reporters about a situation that put them in a negative light. It was vital that they 'come clean' with the facts, but I couldn't convince them. I resigned the account and took a real financial hit. I don't regret the decision, because it was the right thing to do. Sooner or later you'll probably be asked to bend the rules. It's important to adopt a professional code of conduct and know where to draw the line. Your reputation depends on it!"

Lynne Friedmann, public relations consultant,
Solana Beach, California

an engineer doing some tests on a new product and you get the impression that the client wants you to change your report to present the data in a better light, then say something like, "I'm not sure what you're asking. You know I can't change the data, because that would be unethical. These are my test results and this is how I interpret them. Now what exactly are you suggesting?" Often the client will back down when he sees that you are standing firmly on ethical ground. Sometimes, however, the client will be angry or may threaten to take his business elsewhere. Fine—that is not the kind of business you need. Whatever assignments you might lose in the short run will be more than offset in the long run by your honesty and integrity.

> ***"Only once has a client asked me to falsify data. It was a tough situation, because I had sold them on the idea for this particular product in the first place. I thought it was a good idea, but when the tests were performed, the results showed that the device didn't work. I investigated the problems thoroughly and I believed the test results. The customer didn't come right out and demand that I report false results—it was more of an innuendo that I should re-examine the results and report them differently. When I refused, they threatened to take their business elsewhere. That didn't bother me too much, because that kind of business I didn't need."***
>
> **Jack Lewis, engineering consultant, Escondido, California**

Other problems may arise when you have clients who are competitors with each other. This can be a sticky situation if you do assignments for one of the competitors on a regular basis. Most companies will not care at all if you work for a competitor (especially if you do not have access to confidential company information), while a few will be upset at the thought. The best way to handle these cases is to be fully open and upfront about everything with clients and prospects. If you get invited to an initial meeting with a prospect who is a competitor to an existing client, it's a good idea to inform both the prospect and existing client and see if they have any objections. If either has concerns, you can openly discuss areas that trouble them and ways to avoid conflicts of interest or to safeguard confidential information. Of course, you must keep any work you do for competing companies separate. Don't re-use the things you develop for one company in an assignment for a competitor, and don't use confidential information about one client in a project for a competitor. Refuse to discuss any assignment you're doing for one company with anyone at the other company.

Many professions have a written code of ethics. If yours does, it's a good idea to read it and know what's in it. And even if your specialty isn't covered by a code of ethics, reading those of related fields can give you some good ideas on how to conduct yourself. If you always tell clients the truth and keep them fully informed on events affecting assignments you perform for them, you'll go a long way toward avoiding any ethical problems and conflicts.

Angling For More Assignments

The importance of repeat business lies simply in the fact that it costs more money to develop proposals and to seek out new clients. Not only does it cost money, but it takes away from the time you have that you could be charging a client for actual work. When you're selling, you're not making money.

Doing a good job for a client is the best way to obtain repeat business. In addition, when a client is very satisfied with your work, they may do your selling for you. Some of our satisfied technical writing clients have recommended us to their colleagues in other companies—that's an easy way to sell to new clients! But it does no good to patiently wait for more assignments from clients. *Ask them.* Some independent contractors submit a list of ideas for future assignments with their final invoice or submitted materials; a few even go so far as to include a preliminary proposal for additional work. Other contractors subtly feel out clients for future assignments

> **Repeat Business**
>
> ***"75% of my business is repeat business, and the percentage keeps growing. I think the secret is to love what you do. My work is my passion. When I sit down to lay out a book project, the hours fly by and I'm completely oblivious to everything else around me. This level of dedication is what leads to strong word of mouth referrals and repeat business."***
>
> **Sara Patton, book designer and editor, Maui, Hawaii**

during meetings while performing an existing assignment. Whatever technique(s) you use, *ask* to submit more proposals and perform additional assignments!

The Project From Hell: An Actual, Authentic Story

There is no one correct method or formula for handling difficult situations and problem clients. You must consider the client (is the organization incompetent or even dishonest?), possible causes for the problem, whether the client is capable of seeing the problem and helping resolve it, and how badly you want the present and future business from the client. The following account is true, and describes the most difficult project we have yet faced at HighText. It shows what can go wrong and how we handled the situation(s). The company in question is a multibillion-dollar high-technology firm with numerous facilities throughout the United States and the world. All names, cities, and other identifying details have been altered, but all other facts and quotations from letters and memoranda are 100% verbatim. The name of this company does not appear elsewhere in this book or on its back cover. With those caveats, let's begin . . .

We had previously provided technical writing, editing, and document production services for various divisions of a company we'll call SuperTek. We were contacted by another division of SuperTek—which we'll call the flux dynamics division—about doing the production work for a new edition of one of their technical reference manuals. We were happy to get the call from the flux dynamics division, as our previous work for SuperTek had gone well (and paid well) and we were eager to do more work for them.

The initial contact with us was made by "Bob Evans," a vice-president at the flux dynamics division. Bob told us they were putting together a new edition of their *Flux Dynamics Handbook*, a manual of about 300 pages that explained the theory and applications of their flux dynamics products. They had already retained an independent technical writer and editor, "Joe Dean," to prepare the manuscript for the new edition. Joe was a former employee of SuperTek and had worked as an editor for a trade magazine in the flux dynamics field. He had recently left the magazine to become an independent contractor specializing in technical writing and editing. Our tasks were to be relatively straightforward; we were to provide the production services—copyediting, artwork, typesetting, and page layout—necessary to get the book ready for the printer.

SuperTek requested a "kick-off" meeting for this project. The flux dynamics division was located out of state, a few hundred miles away from us, and they would come to visit us. This did not strike us as unusual, since Joe Dean was located in southern California and some of our more distant clients are always looking for a reason to visit us in the San Diego area during winter. Attending the meeting would be Joe Dean, Bob Evans, and "Don Watson," Bob Evans's boss.

The meeting went smoothly. Joe Dean was to be responsible for editorial content and project management, and we would handle only the production of materials received from him. However, we noticed a few unusual things about Dean. He didn't seem enthused about the project. He was not very familiar with personal computers, although SuperTek wanted all text and illustrations prepared using MS-DOS PCs and software. He did not seem to have a firm schedule or plan developed, and there was no logic to his working methods; chapters and sections were not being revised in any discernible order. But since Bob Evans and Don Watson seemed to be sold on Dean, we assumed there was more to him than met the eye. A couple of days after our meeting, we submitted the following proposal to SuperTek:

Re: Flux Dynamics Handbook

Proposed Work: HighText will produce camera-ready copy (defined as resin-coated photomechanical output from a 1250 dpi typesetting machine) for SuperTek's new edition of Flux Dynamics Handbook. All materials for this book will be submitted to HighText by Joe Dean, an editor retained by SuperTek to perform technical writing, editing, development, and project management. HighText will perform copy editing on the manuscript, produce artwork, design the interior pages, set type, lay out pages, and produce an index. All text files are to be submitted by Dean as MS-DOS ASCII files. All original artwork is to be submitted as hand-drawn sketches of sufficient clarity to allow reproduction in final form by a technical artist. HighText will submit all finished text files and illustrations to SuperTek on high density MS-DOS floppy disks.

Specifications:

Trim:	7 x 9⅛ inches
Number of pages:	240
Number of figures:	202
Interior color:	Black only
Cover:	To be designed by SuperTek

For final camera-ready pages:

Editing and page layout:	$4000.00
Indexing:	500.00
Page design:	550.00

Artwork:

128 figures electronically scanned in from previous edition:	320.00
24 figures redrawn from previous edition:	1200.00
30 new figures drawn from Dean's roughs:	1500.00
Typesetting 20 tables:	300.00
Typesetting:	7475.00

Terms: Net 30 days. Individual items above to be billed upon completion.

Schedule: Joe Dean to submit all materials for the revised edition to HighText by May 31. HighText will submit a complete set of page proofs to SuperTek by June 30. Final camera-ready copy to be submitted to SuperTek no later than three weeks after receipt of corrected page proofs from SuperTek.

The project seemed straightforward enough, but it soon became clear that our doubts about Dean were justified. What little material we received from Dean was disorganized and haphazard. At the end of May, we had only received chapters 5, 6, 7, and 9 from Dean. We had performed our copyediting on them, generated the necessary artwork, and released them for typesetting. Dean promised us the remainder of the book would be finished by mid-June. We notified Bob Evans of this situation by phone and by letter. While we were concerned, we decided to wait to see if Dean could deliver the final manuscript by the middle of June before taking further actions.

During the first two weeks of June, it became clear that the project was getting badly off track. The materials we were getting from Dean were disorganized bits and pieces instead of completed chapters. Figures were not labeled correctly, and some figures were submitted twice with different labels. Dean was asking us if he had sent various things to us, as he couldn't remember. The final blow came in the second week of June when he submitted new figures and text changes for chapters 5, 6, 7, and 9—the chapters that were supposedly in final form and which had already been typeset.

We called Bob Evans and informed him of the situation, suggesting that a meeting between everyone involved was urgently needed. A meeting was arranged at SuperTek's headquarters. Dean, Evans, and Watson were to attend. We prepared the following agenda and distributed it at the meeting:

Here are some items we need to address in today's meeting:

1) What is the current status of this project? Which chapters are currently ready to place into production? It had been our understanding that chapters 5, 6, 7, and 9 were complete—in fact, they had already been released for production into page proofs—but now we understand that additional material is coming in for them.

2) Where does the buck stop on this project, i.e., who's the boss? Who says when a chapter is finished and ready to go into production, and make other final decisions relative to this project?

3) What is the most realistic schedule for the rest of this project?

To prevent delays and wasted effort, we strongly suggest that in the future some sort of written authorization be issued by the person in item #2 above in order to release chapters into production or otherwise deviate from the schedule. Copies should be sent to all of us. This would be a good way to keep us all informed and ontrack toward completion in a timely manner.

Finally, our bid for this project was based on an anticipated completion of the final manuscript by May 31 or shortly thereafter. If the manuscript gets larger, involves work other than that specified in our proposal, or is delayed beyond the end of this month, we may have to ask that our bid be altered to reflect these changed circumstances.

The meeting was held on June 19 at SuperTek. We went through the various problems we had encountered. Dean said that maybe he didn't understand what was expected of him, so we went through how the manuscript should be prepared and what our respective responsibilities were for the project. Dean's unfamiliarity with personal computers was painfully evident, and he finally agreed to find an outside service to help him prepare the text files. The delays, incomplete materials, and related problems were discussed, and all Dean could offer was that the project was taking more time than he had anticipated and was more complex than he had first thought. As we listened to Dean's less-than-convincing explanations, we began to suspect that maybe the project was beyond his capabilities.

Bob Evans agreed to be the final authority on all matters regarding the project, including whether a chapter was ready to release into production. Dean and I would send him copies of all correspondence related to the project. However, Evans seemed uninterested, as if the project were not a high priority with him. His boss, Don Watson, asked more questions and acted more concerned.

We discussed the current status of the book. Dean eventually said he could have the manuscript finished by July 31, and we agreed to the new schedule. We warned, however, that any further delays would put us in a real bind as we had several large projects coming up later that year. We received chapters 2 and 4 from Dean, and were told that we could begin the copy-editing and production process for them.

As the meeting ended, Dean asked Evans to approve some additional payments to him for the project. We listened and discovered Dean was billing SuperTek both a fee plus an hourly rate once his hours on a chapter exceeded a specified ceiling. Dean was well over the ceiling on each chapter, yet Evans readily approved the extra payments.

After the meeting, we sent Evans a letter summarizing the key points of the meeting, including the new deadline and working arrangements. But only a few days later the same set of problems started once again. We began receiving material directly from Dean instead of from Evans, and the material was—if anything—more disorganized than ever. Finally, a letter was received from Dean on July 9 that made it clear that something big was still wrong. "Both chapters 2 and 4 are still being reviewed," Dean wrote. "Chapter 4 has undergone considerable expansion and revision to include material originally planned for chapter 15, which is being dropped. I'll give the OK to revise and layout as soon as all reviews are in. You will receive marked-up hard copies with my approval initials. They should start coming the week of July 22."

Needless to say, we were very concerned! Three weeks before, we had been told chapter 4 was finished and ready to go. Evans was supposed to give the final approval on all material, yet Dean was acting as if he had final approval. And the deadline of July 31 was looming and it was clear the project wasn't going to be even close to being ready by then. We realized the project was in serious trouble, and maybe we were too.

Something drastic had to be done. On July 15, we sent the following letter to Bob Evans:

Enclosed is a letter from Joe Dean, our reply to him, and a copy of our notes on chapter status made during our June 19 meeting at SuperTek.

If we sound frustrated in our letter to Joe, it's because we thought the outline and chapter schedule had been finalized back on June 19. (Reference our letter to you of June 20.) It was our clear understanding that chapter 4 was ready to copy edit and that the only additional changes would be those necessary to correct technical errors. We just hope now that chapter 4 won't be extensively reworked to the point that the effort and money we invested is wasted.

As we said back in June, we will have some real problems fulfilling our part of the bargain, both on cost and schedule, if this project is excessively delayed or the final manuscript is larger than the one we bid on. Our original bid was predicated upon having a finished manuscript from Joe on May 31. On June 19, we all agreed that July 31 would be the new manuscript deadline. Clearly—given that we haven't even seen a rough draft yet for 8 of the 15 chapters—there's no way Joe is going to have the manuscript finished by the end of this month or shortly thereafter. Moreover, speaking candidly and confidentially, it's our feeling that Joe has no intention of wrapping up this project anytime soon. There's a real inefficiency in his work methods, and he seems to have a need (financial, emotional, or otherwise) to string this project out as long as he can. After working on hundreds of technical publications projects, we've developed a strong sense for when someone is avoiding completing a project. Based on our experience, we think the earliest Joe can have the manuscript complete and ready for copyediting is this autumn (around October), and that's an optimistic assessment. Given the progress so far (namely, that most completed material is essentially verbatim "pickup" from the previous edition), we wouldn't be surprised if the final manuscript is delayed until next year.

Bob, we really want to do this project for SuperTek, but our bid to SuperTek was based upon certain conditions as described in our proposal. We can't freely and endlessly absorb the time and costs of revisions and delays, nor can we neglect our other clients for the sake of this project. (As we indicated in our June meeting, our schedule for the latter half of this year is heavy.) Hopefully, we will have a clearer idea of where this project stands and the work and time required to complete it by the end of this month. If, as we suspect, the answer is that much more time and effort are necessary to complete it, we will be forced to renegotiate the fee for our services. Should that not be acceptable, or if we can't reach agreement on a new fee, then we will have to bow out of the project and submit a final bill for the services we've performed and costs incurred to date.

This is not the sort of letter we like to write—mainly because we realize we might be slamming a door in our faces—but we're getting painted into a corner by these continuing changes and delays. We thought we all agreed on a final outline and schedule back in June, and it is dismaying, and costly, to see things start to change soon afterwards without any warning . . .

Bob Evans called us as soon as he got the letter. Bob sounded almost mournful on the phone, saying he had worried whether Dean would live up to his promises. (We were slack-jawed on the other end of the phone at this revelation.) Evans confirmed that he, not Dean, had final approval of all materials. He also said that he would immediately talk to Dean and get back to us. When Evans called a few hours later, he said that Dean told him the manuscript would be finished by the first week of August. Dean said he was submitting more manuscript to Evans that very day, and Evans said he would forward the material as soon as he checked and approved it.

After talking to Evans, we realized he was as much a part of the problem as Dean. Clearly, there was no way Dean could have a manuscript ready by early August, yet Evans accepted his promise to do so without question. Further, Evans seemed reluctant to crack down on Dean and force him to stick to schedule. The demeanor and tone of Evans's voice also bothered us; he sounded very much like a man who had already given up trying to make the project work.

Our concern turned into alarm when we started getting "approved" material from Evans. We received the text for chapter 9, but all of the figures for it were missing. Text in other chapters made references to other chapters which didn't exist. On August 1 came the clincher. We received two different copies of chapter 8 from Evans, both of which he had "approved," and the two copies had different and sometimes contradictory changes. Clearly, neither Dean nor Evans was keeping track of what was going on.

We were at a crisis point. The project had turned into a time, energy, and money sink for us, and all of our efforts to correct the problems so far had failed. We resolved that either we were going to abandon the project or receive substantially more money because of the additional work we had done and would need to do to complete the project. The following letter was sent to Bob Evans on August 9:

We like to handle things with humor, and normally we would open a letter such as this with a joke about a "letter bomb." Unfortunately, this subject of this letter is no laughing matter.

We had released the material you submitted back on August 1 for production. Due to problems with it, we had no choice but to pull it out of production. The problems mentioned in our meeting of June 19 and our letter of July 15—namely, disorganization and confusion—are continuing.

We are now at the point where we will abandon this project and cut our losses before they get too much worse unless our agreement with SuperTek is renegotiated to cover the increased work we must do and costs we are being forced to incur.

When we bid on this project on April 11, our non-production services were limited to copyediting, proofreading, and indexing. We did not offer to perform project management or any of the other services Joe Dean was supposed to provide. Yet we found, and still find, ourselves being forced to provide these services in order to keep the project on track. To put it bluntly, this project continues to be disorganized and badly managed.

We are not going to make up for someone else's failures or correct their mistakes for free any longer.

Enclosed are the straws that broke the camel's back. We received from you two copies of chapter 8, both supposedly reviewed and approved by SuperTek, ready for copyediting and production. You'll notice they contain two sets of often different and even contradictory changes.

What are we supposed to do?

This is a perfect example of the disorganization and lack of discipline plaguing this project. Someone should have noticed there were two different versions of chapter 8 and reconciled the differences between them long before sending them to us. This tells us that no one is managing this project and we're expected to make up that shortcoming.

The faxes we've sent over the past few days reflect the lack of overall control. We are missing all figures from chapter 9. We get no instructions on how to handle authorship credit. The text in chapter 2 contains several references to chapter 15, which was supposedly deleted. We could cite other problems that we would had queried about had we not pulled the manuscript from production.

Checking whether the figures for chapter 9 were ready and included, deciding how the table of contents and authorship credit should be given, checking references in chapter 2, etc., are rightfully the responsibilities of the managing editor or project manager, not us.

We have spent far too much time and energy working around someone else's oversights, delays, and confusion.

Our agreement with SuperTek called for us to only do copyediting of the manuscript, proofreading of the page proofs, and indexing. At no time did we agree to put the manuscript together from bits and pieces, manage the project, provide a logic or structure for the project, send chapter copies to outside reviewers, or perform any project management functions. We have done these things because of our desire to continue our working relationship with SuperTek. But enough is enough.

We've "spent" our allocated editorial budget dealing with the problems outlined above. We cannot continue like this any longer. We'd rather walk away now with a painful loss instead of a potentially crippling one later.

Our revised quote for continuing with this project is for our editorial fees to be $9000.00 instead of the original $4000.00 we bid. While the other aspects of our bid remain the same, you should be aware that we strongly expect the changes necessitated to the page proofs for reasons beyond our control will likely drive up the production costs by another 15% to 25%. If this revised bid is not acceptable, we will bill SuperTek for all work completed to date and return all materials to you. We'll still lose money this way, only not as much as if we stick with this project.

We don't like to write letters like this. But we don't like being put in a situation where we have to do so. Please let us hear from you.

This is obviously the sort of letter you write only if you don't really care whether you continue to work on the project or not. But we were at that point.

Bob Evans responded immediately when he received our letter. He did not dispute any of the letter, and placed the blame entirely on Joe Dean ("We've been led down the primrose path by the outside editor we hired"). He agreed to our new terms, and said we would be paid immediately for everything we had completed to date on the project. He said that it would be best to delay any further work on the project until Dean had finished his editing, and if we were busy when Dean completed his editing the project would just "sit" until we could tackle it again.

We were not surprised that Evans immediately agreed to everything we asked for. By this time, nothing about the project surprised us anymore!

Months passed. We heard nothing further from either Evans or Dean. In late November, we received a telephone call from Don Watson asking what the status of the project was. After a short conversation, we realized that Watson knew very little about the problems of the previous summer with Dean and Evans; in fact, Watson thought the manuscript had been completed and we were behind schedule in our production work! He asked for, and we supplied, copies of all correspondence pertaining to the project, including the letters we've quoted in this section.

We next heard from Watson in late January. He told us that he had taken control of the project and we would be dealing directly with him from now on. He also said that Dean had been replaced by another editor, and that we should expect the manuscript in June. It did arrive in June, and was in infinitely better shape than the previous efforts. We were able to finish camera-ready copy in August. The total amount we billed SuperTek was about 70% more than our original bid; this was partly due to the manuscript being longer than the original estimate.

Despite this mess, we have done additional projects for other divisions of SuperTek since then.

Lessons Of The SuperTek Nightmare

We included the SuperTek story because it illustrates some things we did wrong. It also illustrates some things we did right. While hopefully you'll never have a project that's as fouled up as the SuperTek debacle, you'll probably see milder forms of all the problems we had to deal with.

What did we do wrong?

✦ **We didn't include enough detail in our proposal.** We thought it was obvious from our proposal that our role in the project would be limited. Dean and Evans didn't interpret it that way. We should have included a fuller description of what we would do and, perhaps more importantly, what we wouldn't be doing. We should also have spelled out the form we expected all manuscript materials to be in.

✦ **We didn't heed our intuition regarding Joe Dean.** Our initial impression of Dean was that he didn't seem to be the best choice to be the editor for the book. As we'll see in Chapter 8, your intuition can be valuable in deciding whether to accept an assignment or not. We didn't, and we regretted it. We should have voiced our concerns earlier to Evans and/or Watson.

✦ **We let things slide too far before objecting strenuously.** Dean's missing the first deadline should have been a warning shot. By mid-June, it was clear the project was in big difficulty. We should have presented a forceful case then instead of hoping it would somehow turn around.

✦ **We did things that weren't our responsibility.** We performed tasks that were rightly Dean's responsibility in order to keep the project moving. In doing so, we kept both SuperTek and Dean himself insulated from the consequences of his behavior. We left Dean no incentive to improve his performance and encouraged his belief that we would make everything right for him.

✦ **We didn't keep Watson informed.** We should have offered to send Watson copies of all correspondence regarding the project during our June meeting at SuperTek, especially after he seemed to be the one most interested in the project for SuperTek.

What did we do right? Well, there were some things:

✦ **We treated SuperTek as an equal.** Although SuperTek was (and still is!) a vastly bigger company, our letters of July 15 and August 9 had no hint of diffident or subordinate behavior. We did not offer to compromise or volunteer to continue doing the extra work necessary to keep the project moving. We spelled out our case without apology and pulled no punches in placing responsibility for the situation on SuperTek.

✦ **We were willing to walk.** When we threatened to pull out of the project, it wasn't an idle threat. We were willing to cut our losses and quit the project.

✦ **We kept good written records that supported our position.** In addition to the letters to Evans quoted here, we also sent numerous letters to Dean about what he was or wasn't doing. (By contrast, all communications from Dean and Evans were by phone.) There was no way they could claim we hadn't brought problems to their attention or they had misunderstood what we said—everything was in writing for anyone, like Don Watson, to see for themselves.

Many people might object to the way we handled the SuperTek situation. We're not claiming the way we handled the

SuperTek situation was the only correct way or even a good one, but it did work for us. We were able to turn a losing situation into one that was ultimately profitable for us. As an independent contractor, you're going to have your own painful and difficult events. Do as we did with SuperTek—learn from them and apply those lessons in the future.

Chapter 7
Pricing Your Services

Pricing is something that a lot of new independent contractors really dread. In the world of corporate employment, an array of "objective" measures—job classifications, pay ranges, seniority, and so forth—determines how much someone is paid. Such measures are seldom really objective or rational (take a look at the multimillion-dollar paychecks received by executives of money-losing corporations, for example). However, people tend to take comfort in them. As an independent contractor, you can't hide behind a company's pay scales. You'll have to name your price.

Some contractors are so eager to avoid having to put a price on their services that they will ask prospects, "How much do you normally pay for this?" or something equally self-defeating. This is another case of a contractor acting like an employee, and it's especially bad, for it signals to the prospect that the contractor is an amateur. If you do get the assignment, you had better believe that you will be paid substantially less than the maximum the organization would have been willing to pay.

If you're going to prosper as an independent contractor, you must know how to properly set—and defend—a price on the services you perform for an organization. The good news is that the "right price" is probably higher than you might expect.

It's About Value, Not Price

When an organization uses your services, they are almost certainly less concerned with how much your services will cost than the *value* of those services. There are a few organizations that are interested only in price ("the budget says $5000 and that's that!"), but they are few. (They also seem to go out of business with some regularity.)

The concept of value underlies almost all buying decisions. Think about your decision to buy this book. Did you ask yourself "I wonder how much it cost to print that book?" or "How much did they spend on paper and typesetting?" before buying it? The answer is almost certainly no—instead, you were interested in how this book could help you set up a career outside of a company. Your buying decision was based upon whether the information in this book seemed to be worth our asking price, not upon our costs in producing and printing it. That's what the concept of value is all about. And that's how organizations will buy your services. You might feel that you're pressing your luck if you ask a company to pay $10,000 for your services. But if your services result in an additional $25,000 in profit for the company, the company will feel that they got a terrific bargain.

> ***"In my early days as an independent contractor, I got into trouble on some jobs over price. I would try to accommodate customers who had a small budget and give them a 'good deal.' On one particular job I was working for a very low price, and the customer was a loose cannon, wanting to be involved with every little detail and always asking for more and more extra stuff. I finally realized that I was getting clobbered—spending $1000 to make $200! I started to present him with bills for the extra stuff. He got mad, but I stuck to my guns about it. Now I avoid working for customers who argue too much about price initially. I know my prices are competitive and I give good value for the money."***
>
> **Brian McMurdo, graphics artist, Valley Center, California**

A common technique is to price your services on a "costs plus profit" or "costs plus a fixed fee" basis. This means you take your costs for completing an assignment and add in a fixed percentage of total costs, such as 30%, to arrive at the price for your services. The fatal flaw of this method is that it ignores the value the client places on your services and thus may tend to keep your earnings well below what they could be. Another common practice is to price a service at the level of most of the competition. However, the prices charged by your competitors are not always a good indication of the value clients might place on *your* services. What you're offering clients is something unique, something that only you can provide. Why price your services at the level charged by others who not only can't offer what you do but in many cases may offer an inferior version of what you do? If you have done the right job of convincing the client of your ability to handle the assignment, then the fees charged by your competitors will be almost irrelevant, since the client will be convinced that you're the only one right for the job.

Since clients buy your services based upon value, *you will actually lose some assignments if your proposed fee is too low.* This is because prospects will usually assume that something is wrong—you're not really qualified for the assignment, you've overlooked something important or are underestimating costs, etc.—if your fee is too low compared to the value they place upon your services. This actually happens—we have declined proposals submitted to us at HighText because the quoted fee was too low for the value we placed on the services and caused us to have doubts about whether such contractors really knew what they were doing. One of our technical writing clients—a semiconductor company that prides itself on innovation and quality—once told us that they *never* go with the lowest bidder for an assignment, and often accept the highest bid on the principle that "you get what you pay for."

To understand this reasoning, think of situations where you're the client for outside services. Suppose you saw an ad from a dentist offering to cap teeth with porcelain crowns for

$25 per tooth. Would you patronize that dentist? Probably not. You know that most dentists charge a few hundred dollars to make and place a crown on a tooth, and there's no way any dentist can make any money by charging only $25 per crown. Even if the fee were significantly higher, say $200 per crown, you'd recognize that fee was still comparatively low and might have some misgivings about that dentist. The same thing applies to services offered by contractors like you. Organizations that regularly buy services like those you're offering know what the going rate for those services are in your area. If you charge only a fraction of the usual rate, then they are going to look at you as some sort of cut-rate supplier and wonder what's wrong with you.

You'll find that a lot of organizations place a greater value on you and your services when you charge high fees. Again, think of your own experience . . . you probably place a greater value on items you paid a high price for, even if those items are functionally the same as items for which you paid less. Supermarkets are excellent laboratories in which to watch this principle in action. Most supermarkets offer generic products (corn flakes, milk, detergent, etc.) which are identical to higher-priced brand-name products, yet many consumers prefer to spend more on the brand-name items. Why? Many people tend to equate the amount of money spent on an item with the quality and value of the item, a sort of "it costs a lot, so it must be good" rationale.

We all know this is faulty reasoning, but it's something we all fall into at times. Our experience does indicate that you are more likely to be taken seriously and treated like a professional when your prices are high and aggressive. ("They must know what they're doing—look at how much they charge!") In fact, *raising your prices can sometimes bring more work your way.* We had one client for which we performed technical writing and editing services on a retainer basis under a yearly contract. We wanted to end the retainer relationship, but still wanted to keep the client. When our contract came up for renewal, we proposed an increase in our monthly retainer of about 50%. Much to our shock, the client immediately agreed to our increase and started sending even more assignments our way at the higher rates!

Don't forget other elements that determine value when deciding on your fee. Time is a major one. If the prospect needs an assignment completed on a tight schedule and you can comply, that certainly has major value for which you can charge. (You can even explicitly state in your proposal that you're charging for a "rush" schedule.) The amount of competition you have is another. If you have an unusual or esoteric skill (such as a fluency in Arabic), then prospects may find it difficult to locate someone else for the assignment and your very availability has value. Demonstrated experience in successfully handling similar assignments also adds value. (After all, wouldn't you rather spend extra for someone who's shown they can do it right the first time instead of saving some money by using a rookie?)

Of course, there are limits to what you can charge. If a client is used to paying $1000 for certain services, you're going to have a tough time convincing the client you're worth $10,000. However, you can probably get $1500, $2000, or even $2500 for those same services without too much trouble if you can convince the client that you're worth the extra money. To make big money as an independent contractor, you have to work hard and be good at what you do. But you also must be able to look a client in the eye and believe it when you say, "Yes, that's a lot of money, but I'm worth every cent of my fee."

Determining Your Hourly Rate

Some independent contractors bill clients by the hour, while some charge a flat "per-assignment" fee. Others use a combination of these two methods, such as a flat fee plus an hourly charge for any changes to the project requested by the client. Regardless of how you charge your clients, you need to determine what your hourly rate is. Even if you charge a flat fee, knowing your hourly rate will let you know how much you must charge for your time by multiplying your hourly rate by the number of hours you think it will take to complete the assignment. Your hourly rate is the *minimum* you must charge a client above your expenses in order to stay in business; since you price your services according to the value they provide clients, you will often be able to charge more than your hourly rate. But your hourly rate is your "floor price" for your services. If a prospect is unable or unwilling to pay your hourly rate, you don't want to work for them.

The process of determining your hourly rate is simple if you keep a few facts in mind. You should assume that you will be working a typical work week of 40 hours and that you will be taking two weeks of vacation per year. Multiplying this out will give you a total of 2000 working hours per year. (Of course, you can work more hours per year if you desire, but the safe, conservative approach is not to count on any "overtime" when determining your hourly rate.) However, you won't be able to bill all 2000 working hours as an independent contractor to clients. You'll spend time looking for new clients, meeting with prospects, preparing proposals, maintaining and upgrading your skills, meeting with accountants and attorneys, buying supplies, and many other essential tasks not related to anything you do for a client. The actual number of hours you will spend working for clients and charging them will be substantially less, perhaps about 1300 to 1500 hours per year. Those 1300 to 1500 hours are the ones you'll base your hourly rate on. If you can charge for more hours in a year, great—just don't assume you'll be able to do so.

The first step in determining your hourly rate is to figure

out what your overhead will be in a year. Overhead includes such things as supplies, telephone and postage expenses, insurance, travel, equipment, professional association memberships, extra equipment or software, office space, etc., that you'll need to function as an independent contractor. We'll cover the subject of overhead and how to reduce it in the next chapter, but for now let's just assume that your overhead will be $1000 per month, or $12,000 a year. (If you're starting your contracting career, play it safe and overestimate your overhead expenses; anything you can save will be pure profit for you.)

You next have to decide how large a salary you want to earn. Suppose that your salary at your last "permanent" job doing similar work was $40,000 per year. Let's set your desired salary as an independent contractor at $50,000. That's only fair since you have to provide for your own retirement, insurance, etc.

Thus, your goal is to make at least $62,000 (your $50,000 salary plus $12,000 annual overhead) per year after any assignment-related expenses billed to the client. Your hourly rate is $62,000 divided by your total working hours. If you decide that you'll be able to work 1400 billable hours in a year, then your hourly rate will be $62,000/1400 = $44.28, which you round up to $45 or $50 per hour. If you think an assignment will take 100 hours of your time to complete, you would quote a fee of $4500 to $5000 in your proposal plus any necessary expenses.

Keep Your Hourly Rate A Secret!

Some prospects and clients will be curious about your hourly rate, even if you charge on a per-project basis. They may ask you directly what it is, or hint around it. ("How did you determine what your fee would be?") Regardless of how it's asked, we never divulge an hourly rate to a prospect or client unless we are billing by the hour.

One very good reason is that such knowledge can trigger envy and resentment. The client may feel you're trying to take advantage of them—after all, they would never pay one of their employees $50 an hour to do such work! However, the client fails to understand that you have to provide your own insurance, retirement plan, benefits, and pay additional social security taxes out of your hourly rate. Since they don't understand this, the focus of the project shifts from the value you are providing by your services to how much you're getting paid.

An even better reason to keep quiet is that your hourly rate is legitimately a "trade secret" in your contracting business. The client certainly would not disclose such things as the unit manufacturing cost of their products to strangers, and you're under no obligation to provide similar information to the client. Moreover, information about your hourly rate might wind up in the hands of your competitors if the client uses several different contractors.

If you're asked about your rate, try replying like this: "My fee is based upon my estimate of the amount of time needed for this assignment and how difficult the assignment is. My fee reflects typical earnings for someone of my background and experience. Since I have to provide my own insurance, retirement plan, and other benefits provided by employers, my fee includes such items." If the prospect or client persists, politely say that you consider your hourly rate to be proprietary business information—use the analogy to their per-unit costs—and you'd like to keep it confidential. And then politely but firmly dig in your heels.

Handling Objections To Your Price

As mentioned in the last chapter, lowering the fee quoted in your proposal is something you should never do unless you also reduce the scope or type of services you provide.

When you lower your price, you tell the client that you don't believe you're worth the figure you came up with. It says that you're really not that serious about yourself, and the client won't take you seriously either. You also set a bad precedent for all future work with the client, since you will always be expected to reduce your quoted fee in the future. Lowering your fee in return for doing less is a legitimate approach, however; clients often appreciate this approach since it gives them some control over their expenses. But to cut your fee without reducing the services provided is just cutting your throat.

However, what do you do when prospects object to your fee and don't want to reduce the services you provide? Some resistance to your fees is normal and even healthy—if everyone immediately agrees with the fees you propose, you're *under-pricing* your services! The ideal pricing point is where you encounter mild to moderate resistance to your proposed fees and you have no more than 10% of your proposals turned down because your fee is too high. At this point, you're pricing your services at the upper limit of what the market will bear and maximizing your total income (whatever income you lose from rejected proposals will be more than offset by higher income from accepted proposals).

Some resistance to your fees will be "legitimate," meaning the prospect genuinely doesn't feel your services are worth your fee, doesn't understand the reasons for your quoted fee, or simply can't afford your fee. Other resistance will be "illegitimate." You have to know which kind of resistance you're getting in order to respond properly.

Illegitimate resistance is often found when you present a proposal to a group, and is little more than grandstanding by one or more members of the group. Hostile quibbling over minor points or small dollar amounts is a symptom of illegitimate price resistance. Such tactics are used by people who want to demonstrate what tough negotiators they are and how they're carefully protecting the company's money. These people are fond of saying things like "That's outrageous!" or "I know someone who's willing to do that for a lot less!" To handle illegitimate resistance, don't respond with hostility or nitpicking of your own. Instead, be courteous and firm and adopt the position of an expert who has superior insight into how much the assignment will cost. Phrases such as, "Yes, it can be done less expensively, but the quality will suffer," or, "I've had a lot of experience with similar projects, and I don't see how it can be done for less without cutting a lot of corners" will put the quibblers on the spot, forcing them to justify their positions.

Never be embarrassed by your fee or defensive about it. Instead, be confident and even proud that your fee is high: "Yes, my fee is higher than many of my competitors. But my work is also much better than many of my competitors. This is an area where, for good or bad, you do get what you pay for." If you encounter resistance on specific points and dollar amounts, state your reasons why you feel the amounts are justified. Repeat those reasons politely but firmly. Use phrases like, "I know it might seem that way, but in my experience . . ." rather than bluntly disagreeing with assertions made by the prospect. Keep your cool and respond in an authoritative and professional manner.

Be aware, though, that there are some people who *must* get the other person to make concessions in any negotiation. These people feel they must "win" every negotiation, and will not be satisfied until you've backed down on your fee. If you find the prospect repeatedly going over the same points concerning your fee, or sense that the prospect will not stop until you agree to lower your fee, you have to take the initiative and halt the proceedings by saying something similar to this: "I understand and appreciate your concerns about my fee, and I hope I've addressed them. However, my fee is fully justified by the services I would perform and I won't reduce it without a corresponding reduction in the services provided." If you have to say that or some variation of it more than a couple of times, it's time for you to call a halt to the meeting and excuse yourself. You probably won't get the assignment, but such people tend to be trouble throughout an assignment. You're doing yourself a favor by staying out of their way.

The distinguishing characteristics of legitimate resistance are a lack of hostility and a sincere desire to know more about the factors justifying your fee. The prospect often wants to use you and your services, but needs additional information to justify your fee to themselves or their managers. For example, they may not understand the scope or difficulty of the assignment. They may not be aware of all the benefits—and thus the value—of your services to them. Or they may have worked with previous contractors who substantially underpriced their services.

Sometimes the prospect will point to specific items that trouble them ("Your charge for the training sessions seems high") while in other cases the prospect will not be fully aware of exactly what is troubling them ("It seems like a lot, that's all"). In the latter case, "walk" the prospect through each item in your work plan and budget and ask if they are satisfied with the expense of each. Be prepared to explain what is involved to successfully complete each item on the work plan. You should also point out the value each item gives to the organization and how your fee is actually an investment that will help the company function better. Explain what the typical compensation would be for professionals performing those tasks and that someone who is really competent will not work for substantially less than your fee. There will still be efforts to reduce your fee—after all, that's part of their job—but those efforts will be based on facts and logic instead of raw emotion. If you've done your homework in preparing your proposal, you'll be able to get your fee or negotiate a reduction in services for a lower fee.

Sometimes you'll be told that they would love to use you, that you're certainly worth the money you're asking, but they only have so much money budgeted for a project and it would be great if you could do the assignment for the budgeted (i.e., lower) amount. This play for your sympathy and "understanding" is surprisingly common, mainly because too many contractors will go along with it. (We suspect some organizations use this ploy as a way of getting contractors to reduce their fees without directly asking them to do so.) However, their budget has no relationship to how much your services are worth. Budgets are not physical laws of nature; money can be shifted into and out of budgets, and it's been our experience that most organizations will find the money for projects they really want done. If this approach is tried on you, explain that you appreciate their budget restraints but your services are indeed worth the amount you quoted. Ask if your fee could be allocated among different departments or projects to minimize its impact on a single budget. Say your fee can be brought to within the budgeted amount if you do fewer tasks as part of the project. But don't agree to let their budget determine your fee. If it gets down to a take it or leave it situation—"Look, that's all we can pay"—don't be afraid to leave it. You're no different than any other business, like a clothing store or car dealer. If someone can't afford your "merchandise," that's too bad; you'll find someone else who can.

Retainers And Quantity Discounts

If a client is interested in putting you on some sort of monthly retainer or sending a large quantity of work your way, then you can charge a lower fee than you normally would. This is because your normal fee includes your cost of locating new prospects and developing new proposals for their consideration. If a client can save you such costs, you can give a discount out of your savings.

Retainer arrangements vary greatly and are usually negotiated on a case-by-case basis. Typically, retainer contracts involve clients with whom you have a long-standing relationship. You both understand each other and the client understands how much you will charge for an assignment of a given length and complexity. Since the relationship is ongoing and somewhat "co-dependent," a retainer contract allows the client to secure a commitment of a certain amount of your time each month on an exclusive basis and you in turn are guaranteed a certain amount of income each month. For example, you might receive a monthly retainer of $2000 from a client. If you bill at $50 per hour, this would mean that you would allocate 40 hours each month for the exclusive use of that client. If you bill on a per-project basis, you would block out enough time in your work schedule to do a "$2000 project" for the client. The key is that you would get that $2000 each month even if you did no work whatsoever for that client. In return, you would give the client's projects absolute priority over all other assignments until you had "earned out" the retainer fee.

(As a practical matter, you would continue to give that client's projects priority to the maximum extent you possibly could.) Any work done in a month above that covered by the retainer would be billed at your customary rates.

One pitfall of a retainer is that you can become too dependent on a monthly retainer and not develop enough additional clients. Another problem is how to handle any unearned retainer payments. Many clients will want to roll forward any unearned retainer fees. For example, if you're paid $2000 per month but only do $1000 of work in a given month, most clients will want you to do $3000 of work the next month before paying anything additional. The problem for you in such an arrangement is that the client might let the unearned hours accumulate and suddenly hit you with a flood of work in a month, possibly causing you to neglect other clients.

Possible solutions to this include setting a cap on the maximum hours you'll work for the client in a month, putting an expiration date on the use of unearned retainer fees (for example, retainer fees could not be carried forward more than two months), or reducing the amount of unearned fees that can be

"I've found that retainers have pros and cons. The advantage is cash flow management. Also, it's easier to 'drop in' on clients for a chat or a cup of coffee without the client worrying about the 'meter running.' This access helps to advance ongoing projects and identify new ones. The down side to retainers is client expectations. For some reason, clients seldom think they're getting their money's worth with a monthly retainer. In addition to meetings and project updates, at least twice a week I mail something to my clients, whether it's a news clipping, a note, whatever . . . the more informal the better. It helps to solidify the relationship and lets them know I'm 'on the job.' "

Lynne Friedmann, public relations consultant,
Solana Beach, California

Friedmann
Communications...

P.O. Box 1725 • Solana Beach, California 92075 • (619) 793-3537

April 14, 1992

Carol S. Lewis, President
HighText Publications Inc.

Dear Carol:

This letter, when signed by you, will confirm that HighText Publications Inc. has retained Lynne Friedmann as a public relations consultant for three months, beginning May 1, 1992.

As HighText's public relations counsel, I will develop for your approval and implementation a full service program designed to support the promotion and marketing of "Hands On Science" in addition to pursuing specific PR opportunities designed to raise the awareness of HighText among the local technical community. For these services, HighText agrees to pay a minimum monthly retainer of billed on the first of each month in advance, and due and payable in ten days.

I will work against this retainer at an hourly rate of If program activities for any month should require the expenditure of more time than the retainer covers, I may seek your approval to bill additional hours at the same hourly rate of

HighText will reimburse out-of-pocket expenses incurred on its behalf for items such as postage, facsimile transmissions, long-distance telephone calls, mileage, etc. Extraordinary expenses, such as for travel and entertainment, will be incurred only with your prior approval. Expenses will be rebilled at net cost with no markup.

Along with a monthly invoice, HighText will receive a detailed report of activities performed and results obtained.

This agreement may be modified in writing only. Either party may terminate this agreement with 30-days' written notice.

Figure 7-1: An actual retainer agreement HighText entered into with Lynne Friedmann, an independent contractor specializing in public relations, marketing communications, and publicity.

(continued on next page)

Carol S. Lewis, HighText Publications Inc.
April 14, 1992
page two

Please retain the signed original of this agreement for your records and return a signed copy.

I look forward to a long and mutually beneficial relationship with HighText.

Sincerely,

Lynne Friedmann, Consultant

ACCEPTED AND APPROVED
HIGHTEXT PUBLICATIONS INC.

BY: ______________________________, President

DATE: ____________________________

Figure 7-1 ***(continued)***

carried forward and setting a cap on the total (for example, only half of an unearned fee could be carried forward from a month, and then only up to a certain total). Retainer agreements can be a great way to smooth out your monthly cash flow and plan your work schedule better, but they must be structured carefully to avoid flooding you with work at inopportune times.

Quantity discounts can be introduced during the negotiation stage of your proposal. If the client wants you to do more tasks over a longer period than you originally proposed, you can offer a nominal discount—say 5% or 10%—from what you would charge for the same tasks with a new client. You can do the same with the fees in future proposals. However, don't discount your fee in your original proposal in return for a vague promise to send future work your way; discount your fee only for firm assignments. If the client is serious about additional projects for you, they may be willing to consider a retainer contract with you.

Setting a price for your services is no place to be faint of heart. If you have to make an error here, you'll probably be happier and make more money if you overprice rather than underprice your services!

Chapter 8
Random Thoughts

This chapter is a "grab bag" of ideas, thoughts, and observations left over after writing the preceding chapters. These are some of the principles, tips, and tricks we've tried and found to work. Some of them may not be applicable to your situation, but hopefully they'll start you thinking.

When Is A "Good Time" To Become An Independent Contractor?

A large percentage—maybe even a clear majority—of those who read this book will never even attempt to become independent contractors despite their interest in doing so. Most will have the skills and energy to succeed on their own, but they won't even try. If you were to ask them why they don't, the most common response would probably be some variation of "Now's not a good time." On the surface, their reasons would seem quite sensible . . . they want to get the car paid off before striking out on their own . . . they want to wrap up this project at work that will really impress potential clients . . . they haven't decided precisely what they would like to do on a contract basis . . . and so on.

Guess what? *There is never a "good time" to become an independent contractor.* No independent contractor has ever started with everything exactly in order the way he or she wanted it. All of us started out with risk and uncertainty staring us in the face. We all had bills to pay in the future and the possibility of getting sick without a lot of money in the bank. However, this situation is not much different from "permanent" employees in this era of downsizing, restructuring, mergers, and other elements of widespread corporate implosions. Embryonic independent contractors tend to be more aware of the risks facing them, however, and are in a better position to do something positive about them than permanent employees who rely on the beneficence of corporate executives.

If you're waiting until everything's "right" before trying to become an independent contractor, you'll be waiting forever.

It's possible to test the waters for independent contracting while you're still a permanent employee, however. You can compile lists of prospects at night or on weekends. You can make some calls to get names during your lunch hour or on a day off. You can even set up meetings for vacation or personal days when you're not at work. If you're uncertain whether independent contracting is for you or whether there are enough prospects in your area to make it worthwhile, this is a low-risk way to test the water. But eventually you'll have to jump in head first if you want to be an independent contractor. And today is as good a time as any.

> ***"At first I thought that I wanted to be an independent real estate agent. I promised myself that I could leave my regular job and pursue that goal once I had saved enough money to survive for one year with no additional income. I strongly wanted to work at something I loved, with money secondary to satisfaction and enjoyment. Once I had the money saved, that gave me the freedom to look at what I really wanted to do, and I discovered that book design and desktop publishing were my true loves, not real estate! But it seems that achieving that first goal—giving myself the space of a year to examine what it was I really wanted to do—was an essential step along the path."***
>
> **Sara Patton, book designer and desktop publisher, Maui, Hawaii**

Avoid Overdependence On A Single Client

It's not uncommon for independent contractors to eventually land a big contract—representing over half of their income—with a single client. They work long and hard at that contract. Everything is fine until the contract is completed and their remaining clients, if any, are not enough to keep them going. A variation of this often occurs when a senior manager or executive is terminated and the company gives the ex-employee a "consulting" agreement that lasts a year or two. At the end of that period, the ex-employee has no other clients and is twisting in the wind.

When you land a big assignment, it's tempting to think that you've got it made for a while and slack off on your efforts to locate new prospects and turn them into clients. Yet all contracts come to an end sometime, and you'll need new clients and new assignments to keep going. That's why you have to avoid becoming overly dependent on a single client, no matter how lucrative your arrangement with them might be. The exact "danger point" that indicates you're overly dependent on one client varies according to your field and services offered, but any time a single client represents 25% or more of your income you have cause for real concern. If one client provides more than half of your income, you're no longer a true independent contractor; you're a *dependent* contractor!

It's also easy to slip into an employee mindset if one client controls a major chunk of your income. It's hard to tell such a client "no" even if a request is patently unreasonable. You're in a poor negotiating position if the contract comes up for renewal; how can you afford to risk losing such a big contract if you don't have others ready to use your services? And, despite your best efforts, you may find yourself neglecting your smaller clients in order to service the big one. Your smaller clients will eventually find others who give their assignments the attention they deserve, and you become even more dependent on that big contract. You might even find that the IRS considers you to be an "employee" of the company that gave you the big contract.

If you do land a big, lucrative contract, great! But that's also a sign to redouble your efforts to find other clients and assignments. Big contracts come and go; the only thing you can keep constant is your effort to expand and upgrade your client base. A few phone calls and letters each week is all it takes to make sure you're not left stranded when that big contract expires—and it will expire some day.

Leave Some Slack

You need to leave some empty time in your weekly work schedule. A good rule of thumb is to have about 20% of your work week open and uncommitted whenever possible.

There are some good reasons for this, particularly if you have a group of clients that you work for on a regular basis. Those clients will have emergencies and rush jobs that will demand your urgent attention. If you don't have some slack in your schedule, you won't be able to help them and they may have to find other contractors who can. You're also human; some things will take longer or be more difficult than you anticipated in your proposal. You might get sick, break a leg, need to get your car fixed, etc. And how will you be able to quickly jump onto a lucrative new assignment or add a new client if you have no free time in your schedule? Even if you don't have an emergency or new assignments, you can still use your free time to look for new clients or improve your skills.

Remember, your goal as an independent contractor is to maximize your income, not the number of hours you work in a week. But some independent contractors seem to feel secure only when they have every waking hour booked with assignments. They wind up painting themselves in a corner, as they find themselves without the time to handle emergencies from their regular clients or to accept assignments from new clients. It seems paradoxical, but it's true: leaving some open, uncommitted time in your work week is a great way to maximize your income.

Why Clients Disappear

Your meetings with a new client had been promising, with some broad hints of additional projects waiting in the wings once your current one was completed. You wrapped up the assignment, collected your final check, and called from time to time offering to meet with them and develop new proposals. But their enthusiasm seems to have diminished since your early meetings; they give only vague reasons for not wanting to meet with you or consider additional proposals. Eventually you realize they simply don't want to work with you again, and that something happened during the project to make them sour on you.

Unless you do something horrifically wrong, clients will seldom tell you if or why they are unhappy, even if you ask them directly. Instead, they'll tell you they have nothing for you now or that those other projects you discussed have been postponed or their budgets have been cut.

One reason clients might disappear is—to put it bluntly—incompetence on your part. If you botch a project, take longer than you promised, or do something other than what you promised in your proposal, then you're not going to get a second chance. But you'll know if you were in over your head or exceeded the schedule. What about those situations where everything seemed to go well from your perspective, but something must have upset the client?

As we said earlier in this book, trust is the most important element when companies decide to use an independent contractor. It's also the most important element when organizations decide not to use a contractor for additional projects. Even if an assignment was completed successfully, the company may feel like they dodged a bullet and to use you on another project would just tempt fate.

Trust can be shattered in any number of ways. Your total image counts for a lot. No one ever thinks they're rude . . . arrogant . . . short-tempered . . . snobbish . . . or a jerk, but a lot of people are exactly those things. Some independent contractors put on their party manners for the initial meeting or presentation, and then revert to their normal crabby selves once their proposal is accepted. But you're being evaluated each time you come into contact with someone from the client company. It only takes one incident to find yourself *persona non grata* at an organization.

People will distrust you if you don't listen to them. From our own experience, few things are more frustrating than telling an independent contractor what you want done—and then getting back something not even remotely like what you described. It's not that such contractors deliberately ignore what we tell them; they're so focused on the task during our discussion that our comments just don't register with them. But regardless of the motive, we're disappointed and think twice before giving them another assignment.

One thing that is always fatal to trust is dishonesty. If you lie to or mislead a client over even a small matter and the client finds out you've lied, they will wonder if you've lied to them over bigger matters. You don't have to lie over a big matter (like your previous job experience) to hurt yourself irreparably; even a "white lie" like blaming someone else—a printer, a shipper, or a supplier, for example—for a delay in completing a project can destroy your credibility if the client finds out it's not true.

Withholding information from a client can be almost as damaging as a lie. You might be reluctant to tell a client about a situation that might delay the project or alter the outcome, and decide to wait to see if things will work themselves out. Or you may find some bad news within the organization—like a quality control or personnel problem—that you suspect the client might not want to hear. Regardless of your misgivings, the client is entitled to know when something is not going according to plan as soon as you're aware of it. If the client later becomes aware that you knew of the situation but kept it to yourself, the client will rightfully be angry at you.

Clients can also become disenchanted with a general lack of professionalism on your part. Small mistakes and errors that seem minor at the time can add up to produce a negative impression of you even if an assignment is successfully completed. If a client feels he or she has to supervise you and double-check your work every step of the way, then they may feel they'd be better off to hire a permanent employee for the task. And your attitude can work against you. No one likes a whiner, complainer, or someone who gives the impression they'd rather be somewhere else, and those attitudes are especially resented in independent contractors.

Every contact you have with a client after acceptance of your proposal is a chance to show them that they made the right decision in using you. The same enthusiasm and attentiveness to the client's needs that you used to land the assignment are also needed throughout the assignment. To win a client's trust and confidence is not easy and takes time. To lose a client's trust and confidence is easy and takes only seconds.

Plan, But Don't Overplan

You do need to do some planning as an independent consultant. Of particular importance is some form of cash flow projection. This is a monthly projection of your expenses for a given month along with your projected income. Having some idea of how much cash you will need at certain times is crucial, since a cash shortage can cripple or even end your independent contracting career. You also need to develop some general goals for each year (like adding six new clients and increasing your net income by 15%) along with a strategy and tactics for reaching those goals.

But don't overdo it and develop an overly complex business plan. Big companies spend a lot of time and effort on their planning functions, and they usually miss by a mile. Take a look at well-known corporate horror stories of the past decade—IBM, Digital Equipment, Pan American Airlines, Wang, Fairchild Semiconductor, etc.—and you'll notice all were companies that had large numbers of employees developing plans for them. So what makes you think you are any more psychic than they were?

Moreover, extensive planning works against one of your biggest strengths as an independent contractor. Once a big company sets out along a certain path, it's difficult for it to change directions if circumstances abruptly change. (For example, look what happened in the domestic auto industry when increased oil prices created a demand for smaller cars in the 1970s.) You, however, can adapt to different conditions and profit from them. Independent contractors are usually more up-to-date on new developments, technologies, and tools than most permanent employees, so rapid changes in your field can actually increase your attractiveness to organizations. But if you insist on rigidly following a plan, you might miss these opportunities.

Stay flexible and adaptable—plan, but don't overplan.

Culling Clients

Your goal should always be to improve the quality of your clients—that is, how well they pay and how interesting their assignments are. This means that each year you should make it a goal to replace your "bottom" 15% or 20% of your client base with new clients. The obvious way to do this is to simply stop submitting proposals or tell them you're too busy with projects already underway to take on additional work. But if the major problem with a client is low pay, a better approach is to drastically raise your rates to see if the client will drop you. You might be pleasantly surprised at how readily the client agrees to pay your higher fees. In one case, we wanted to decline additional work from a client who had been difficult to work with because of their internal disorganization. When asked to submit a bid on an upcoming project, we raised our fee 50% from the last project we had completed—and our bid was quickly accepted! And, as we described in the last chapter, a similar thing happened for a client we had a retainer agreement with. If the client balks at your higher fees, or if the problem involves something that can't be compensated for with more money, then find another client. But aggressive pricing with a client you were going to drop because of low fees can sometimes pay off.

Your Overhead

Needless overhead has brought down countless independent contractors. Every expense not directly related to producing income is suspect.

This doesn't mean you should scrimp on legitimate expenses or the tools you'll use in your work. It does mean you should ask yourself whether an expense is really necessary. It's easy to fall into the trap of thinking a problem will be solved if you just have a more powerful computer . . . buy a new software package . . . or attend a trade show on the other side of the country. You might be right to think that way, but you might also be wrong. One good analytical tool you can borrow from production management people is the concept of a "payback" period. The payback involves estimating how long it will take for an expense (such as a new computer, software, office furniture, attending a convention, etc.) to be earned back through additional sales or profit resulting from making that expenditure.

One big question is whether you should have a separate office or work out of your home. If your services are the sort where clients expect to come to an office, then investment in a separate office space is certainly worthwhile. In some cases, having a home office—even if it's in a clearly separate space from the rest of the house—will cause clients to think of you as less than professional. However, if you normally visit your clients for meetings, then you can get by fine with a home office. If you do use a home office, however, it's best to have a separate business telephone and fax/modem lines. Not only does this make keeping track of your expenses easier, it also prevents missing a call from a client because someone else was using the telephone.

One consideration is favor of separate office space is that the IRS has become much tougher in its rules for deducting home office expenses. Having a home office is now almost an open invitation for a tax audit. A separate office also helps you meet one of the legal tests for being an independent contractor. Finally, you might find it psychologically more comfortable having an office strictly for work. We've both tried home offices, and found it hard to keep our business and personal lives separate when we lived where we worked. This might not be a problem for you, of course, but is a factor to consider.

Partners

HighText is a company owned by two partners—the authors of this book. Despite some heated disputes and arguments over the years (a few trash cans have been kicked considerable distances, for example), we do serve as good "reality checks" on each other and some real synergy results from being partners. Having said that, we urge anyone thinking of entering independent contracting with a partner or partners to be extremely careful in evaluating whether such a relationship is workable or beneficial. It's important to have a written agreement governing the partner relationship and that some method of resolving disputes and ending the partnership be spelled out in the written agreement. The reason why we feel so strongly about this is that we had an additional partner when we founded HighText, but that person left ten months after our founding. Fortunately, HighText's corporate by-laws included procedures for buying out the shares of any partners and formulas based on the company's net worth for determining the buy-out price. A very messy situation could have resulted had we not had such agreements in place.

You must first decide *why* there should be a partnership. Personal reasons, such as a friendship among the partners, are among the most common reasons for people to work as partners. However, these are sometimes the worst reasons for a partnership. You're starting a business, not a social club. Close personal relationships may actually make it difficult for partners to honestly disagree or speak out when things go wrong, since they may be more concerned with their feelings for each other than they are in trying to make the business relationship work.

A good partnership requires agreement on fundamentals, such as business goals and operating principles, yet needs enough variety in skills and attributes so that the partners can together accomplish more than their sum total as individuals. Many good partnerships are based on contrasting skills or attitudes among the partners, as when one partner is cool and analytical while the other is high energy and creative. Another good mix occurs when some partners have technical, engineering, and/or production skills while the others have marketing and financial backgrounds. It's crucial to determine how decisions will be made and disputes resolved; methods can include majority vote, electing a "managing partner" who will act as a sort of chief executive for the partnership, and mediation or binding arbitration. None of these are a substitute for having partners who are willing to compromise, swallow their ego, and put the interests of the partnership as a whole ahead of their individual interests. You can never be sure whether a partner has such abilities until the first serious dispute of the partnership.

Partnerships formed with co-workers from your previous employer have a hidden danger. Departmental rivalries and personal relationships between employees are often carried over into the partnership. If the "engineering guys" in a company hate the "financial guys" and vice-versa, then it's a good bet that the same thing will be true in a partnership. Moreover, too many people from the same company can create a situation where the partners attempt to re-create that company instead of developing a real partnership of independent contractors.

Our advice here is hypocritical since we have a partnership that works, but our suggestion is generally to avoid partnership arrangements unless you know your potential partner(s) well and there are compelling business reasons to be partners.

Networking

As an independent contractor, you probably won't hire permanent employees as such, but situations may arise when you need to have tasks performed that are outside your area of expertise. Who do you turn to? Other independent contractors, of course! Forming informal networks with other independent contractors can greatly enhance your income potential. These networks are essentially "virtual corporations" that are created to complete a certain task and then go out of existence. Networking with others often makes more sense than becoming partners with them.

HighText has informal relationships with independent contractors specializing in computer programming, electronics engineering, multimedia development, technical illustration, book production, commercial art, printing, and similar fields. These relationships are mutually beneficial to all of us. When we get a major technical manual project, we farm out parts of it—like illustration or page layout—to other contractors who specialize in those areas and can do them more cost effectively than we can. And in turn these independent contractors send work our way, as when the computer programmers ask us to develop the manual for software they are creating for a client. We share information about prospects with each other ("They look like they might need some technical illustration for their internal documents") and offer support and advice for each other during rough periods.

Some areas have formal organizations of independent contractors, usually by field of specialization. More often, however, you will form relationships with other contractors you come across in your normal activities. Other good places to locate other contractors are on computer bulletin boards or on-line services that have special interest groups for the "self-employed" or those who work at home.

Trust Your Intuition—Another Case Study

We've mentioned previously in this book how intuition can be valuable to an independent contractor. We're not saying to avoid rational analysis, but we are saying that you shouldn't ignore a warning voice in your head if you hear one. Here's an actual example illustrating how intuition can work . . .

A few years ago we were contacted by a large software company and asked if we were interested in bidding on a large manual project they had coming up. We had contacted this company earlier, but were told that they did all of their manuals with an in-house staff. This company was a recognized leader, so we were excited at the possibility of doing a manual for them.

At the initial meeting, we had to sign a confidentiality agreement and were shown the software package under development. It was a highly advanced computer graphics system, and we were impressed; we remarked that we needed to buy it when it was released. We had a lengthy discussion afterward with the company's director of technical publications. He stressed how important this product was to his company, and that the schedule for producing the manual was tight. Our meeting took place in July, and the manual had to be complete and ready for the printer by the end of December, since the product would be shipping in the first quarter of the next year. The manual would be large—from 250 to 300 pages—and would be both a tutorial and reference.

From the questions we asked during this meeting, we learned the software package was far from being in final form; in fact, it wasn't yet at the "alpha" stage (the first working version submitted for independent testing). We knew from experience that it's difficult to make a realistic estimate of the schedule and fee for a software manual project unless we know what the final form of the software will be. We told him we could make a tentative estimate based upon the projected manual length and similar projects we had worked on in the past. However, we would need a copy of the final design specifications for the software before we could make a firm estimate. The final "specs" would give us a concise overview of the software package and how difficult it would be to describe it to someone else in the manual. We agreed to meet again in a couple of weeks. At that meeting, we were promised a copy of the final design specifications. We would also meet with a couple of their in-house technical writers.

Discussing our meeting later, we realized this would be a major project that would require numerous "overtime" hours on our part to meet their schedule (we would have to master the software package before we could write about it). It would place a major strain on us and our resources, but felt it would be worthwhile if we could earn enough from it. We decided to be very aggressive on our fee, substantially above what we would normally charge. However, we felt it was justified by the schedule. Moreover, the company was a successful one capable of paying well.

Our next meeting was with the director of technical publications and two of his writers. Our high fee didn't raise an eyebrow. We were very impressed by the two technical writers we met and developed an immediate rapport with them. In fact, the writers seemed enthusiastic about the prospect of us doing the manual instead of them. We discussed the status of the software package, and learned a lot of frantic effort was being made by their programming staff to get it finished by the end of December. The final design specifications weren't ready yet, but the programmers assured them they would be finished soon. At then end of the meeting, we agreed to submit a final proposal with a final fee quotation within five days. We left with smiles, firm handshakes, and expressions from them of how much they'd like to use us for additional projects if this one worked out well.

As soon as we left the entrance to the company's building, we started discussing what our response would be. Before we had walked the 200 feet to our parked car, we had agreed that we didn't want to do the project.

If you're puzzled by our behavior, so was the company. When we phoned their director of technical publications to tell him we were declining to submit a final proposal, he became angry. His anger was compounded by the fact that we couldn't offer any specific reason for our refusal. We simply said that it seemed the project was already off schedule and that it would be difficult, if not impossible, for us or anyone else to have a manual ready to go to the printer by the end of December. He forcefully said the project was not behind schedule, that he would find someone who was capable of meeting the December deadline, and abruptly hung up.

It turns out we were on to something. The software did not ship in the first quarter of the next year as we were told. In fact, it didn't reach market until almost 18 months later. So how did we know the project was troubled? One very big clue was the lack of any firm design specifications at such a late stage of development. In most large software projects, design specifications are completed before a single line of programming is done. If even preliminary design specifications weren't available this late in the game, something was clearly wrong. Without design specifications, the various programmers working on the project would be going off in different directions. That would compound the number of errors and bugs in the software, and we couldn't see any way the project could stay on schedule without such specifications.

Moreover, the two technical writers were clearly competent individuals. They had long been familiar with the software; they could start writing the very next day, whereas we would have to learn the software before we could write about it. If making the December deadline was really that important, it would have made far more sense for the company to use their in-house staff and let us handle something not as time sensitive. The two technical writers were more than professionally courteous to us; they seemed almost overjoyed to see that someone else was going to handle the project. And there was no resistance to or even discussion about our fee, which was clearly high.

Rolling it all over in our minds, we concluded they were desperate to unload the project on someone because the project was in trouble, there was no way the December deadline was going to be made, and they didn't want to be stuck with it.

While we had no further contact with anyone at the company, the 18-month delay in getting the software to market showed that our feelings that the project was in trouble were correct. Had we gone ahead with the project, we would have been stuck with a situation as bad as—or maybe even worse—than the SuperTek mess discussed back in Chapter 6.

We're certain that we have missed out on some good assignments by relying on our intuitive feelings. But every time—without exception—we have accepted an assignment contrary to our intuitive feelings against taking it, we have had major problems. If you sense something is wrong with a project even though all the rational, quantifiable factors are favorable, listen closely to that voice in your head. You don't always have to heed it, but you had better not ignore it.

Getting Exposure

In addition to your normal prospecting for new clients, you should also make efforts to establish yourself as an expert in your field. Common ways to do this include speaking at seminars and trade shows, writing articles for professional journals covering your field, becoming active in trade groups and associations in your area of expertise, making yourself available to newspapers and other media as a source in your field of expertise, doing radio and television interviews, or publishing a newsletter. There isn't enough room in this book to tell you how to do such things, so instead we recommend you read one of the many books available on how to conduct a "self-publicity" campaign.

From our experience, it's almost ridiculously easy to promote yourself, since the different media have a voracious appetite for new stories and people to interview. For example, there are biweekly publications for newspaper reporters and radio/TV talk show hosts. A simple, relatively inexpensive ad in one of those publications can result in several print stories about you and radio or television interviews in which you can promote yourself and your services. Making yourself into an expert will result in some clients seeking you out and may even let you start charging higher fees. After all, you will be an expert—you'll have a long string of newspaper stories and radio/TV interviews to prove it!

Offload The Trivial Stuff

As an independent contractor, your time is the most precious resource you will have. You can figure out how to make more money or you can buy more equipment, but those 24 hours in a day are all you will ever have. And that's why it's amazing how many independent contractors fill their days doing trivial tasks that steal time from what should be the main business of their contracting career. These contractors rationalize their behavior as saving money. But by taking time away from their income-producing activities to do low-level maintenance tasks, they're cutting their income more than they're saving on expenses. Would you pay someone $40 or $50 per hour to do some simple bookkeeping or just to copy things for you? The answer is certainly not. Yet too many independent consultants who bill their services at those rates will spend hours each week doing tasks they could pay some other independent contractor or service to do at substantially lower rates.

Saving money is great, but that's not your main goal as an independent contractor. You're in business to maximize profit, which is your total income less your total expenses. If the choice is between doing billable work at $40 to $50 per hour or doing work yourself that someone else is willing to do for $10 to $15, it's doesn't take a genius to see which course of action makes sense. Yet many contractors are so intent on saving money any way possible that they lose track of these simple realities. Or they may tell themselves that they had nothing to do anyway, so why not save a little money and do the stuff themselves? But that's still self-defeating. As an independent contractor, no working hour has to be idle. If you're not completing an assignment for a client, you need to be looking for new clients or improving your skills to better serve clients.

While it's normal to do housekeeping and maintenance tasks yourself in the beginning, your contracting business will grow to a point where you need to decide what you do that is unique and has value and what you can safely unload on service agencies or other contractors. When we started High-Text, we did everything ourselves, even down to cleaning our office. As we grew, our accounting and bookkeeping functions were the first to go outside, mainly because we did not feel we had an adequate background in those areas. Technical artwork was next, again because our skills there were not exceptional. As our work load grew, we discovered that paying others to do page layout and design freed us to spend time doing the more lucrative technical writing and editing. Eventually we reached the point where most of those tasks we did for ourselves in the beginning are now done for us by others. Our time is now spent directly on those activities that generate the most income per hour of effort.

Some independent contractors rationalize a "do it all" approach by saying they want to keep control of all aspects of their business. However, this makes sense only if doing everything adds value to your services that customers are willing to pay for. In some rare circumstances, doing everything might enable you to make deadlines more reliably or assure a high level of quality. But these cases are rare, and will work only if you can perform all of the tasks to a high level of competence. Maybe you are someone who really is simultaneously a talented graphics artist . . . computer programmer . . . accountant . . . printer . . . PC technician . . . housekeeper, and can do it all. But you're probably not. If you fall short in an area that's important to your clients and still try to do everything yourself, you might wind up with ex-clients instead. If you fall short in other areas, such as accounting and bookkeeping, you might find yourself with severe tax or financial problems.

Don't be a control freak. Find out what activities generate the most profit for you and focus on doing them. Find others to do the remainder.

Avoid Righteousness And Creeping Elegance

It's good to want to do the very best work you possibly can on each task you perform as an independent contractor. It's even better to realize that sometimes clients just want an adequate job instead of your very best.

Creative types—like artists, writers, and engineers—are especially prone to this syndrome, which we call "creeping elegance." As they work on a project, they gradually add more "improvements" that result in the end product being quite different from what the client anticipates. A client might want a simple, straightforward drawing or plainly written, business-like text. Instead, he or she gets a "work of art" that might win awards from other artists and writers but is not what the client wanted. When the client objects, the contractor delivers a lecture on how the client is mistaken and obviously unable to appreciate professional quality work. In other words, the contractor gets "righteous" with the client. Contractors in other fields do the same thing. Accountants set up accounting systems that are too complex for clients to understand, marketing specialists submit marketing plans too expensive for clients to implement, engineers design products whose exacting specifications are too difficult for a client company to meet in actual production . . . and so it goes.

Creeping elegance and righteousness are caused by forgetting that you're hired by a client to solve a problem, not demonstrate how clever you are. Clients may not know what is "professional" or "classy" in your area of expertise, but they know what they like. And if you don't give clients what they like, they won't care how professional or classy your work is. This is especially true if you're billing by the hour and an assignment winds up being more expensive than the client expects in addition to not being what they wanted.

Sometimes, all a client wants, and is willing to pay for, is an adequate solution to a problem instead of the best solution. If a client tells you they want a "quick and dirty" solution instead of a fancy one, give that to the client. Should you genuinely feel that a simpler, less complex solution desired by a client is wrong, voice your concerns *before* completing (or accepting) the assignment. If the client insists, and you strongly feel that it is wrong, then decline or gracefully bow out of the assignment.

Guarantee Your Work

We offer prospects a 100% guarantee of satisfaction in all of our proposals. If the client is not completely happy with our work, they owe us nothing. That's not as big a risk as it might seem, for if clients aren't happy with our work, we probably won't get paid anyway. The same will apply to you. If you don't deliver on your promises, the client might pay only a fraction of your fee or even refuse to pay you at all.

Unlike a permanent employee, you're not being paid to "try." You're being paid for results even if you bill by the hour. Think about it from the times when you've been a "client." Would you pay without question the bill of an auto mechanic who had failed to repair your car? A plumber that couldn't stop a leaky faucet? A painter who paints half of your house and then walks away from the job? The answer is clearly no, but some new independent contractors expect to be paid for making an effort even if they don't produce the results they promise. But clients are probably well within their rights to refuse to pay or pay less than the proposals calls for if you fail to do what you promised. You can threaten to sue (or, more likely, take clients to small claims court) if this happens. But a court may well rule against you if you didn't deliver in all respects on what you promised in *your* proposal.

So far at HighText, we've been fortunate enough not to have any of our clients refuse to pay or pay only a portion of one of our bills. But we have reduced payments to independent contractors we have retained and have refused to pay a couple of invoices. One case involved a defective computer hard disk sent to a "data recovery" specialist. This person was able to recover many of the files on the defective hard drive and put them on a good hard disk we had provided. However, this person's computer system was infected by a computer "virus," and the recovered files we received from him were therefore infected by the virus. Since our computers were all networked together, all of our computers were rapidly infected. We were able to determine that a virus was responsible and that the "infected" file was on the hard disk received from the data recovery service. We had already paid the data recovery service fee (which was a few hundred dollars), but stopped payment of our check immediately when we discovered the virus. Amazingly, the data recovery specialist still wanted to be paid even after we told him what we had done and why; he argued that he had, after all, recovered most of the files like he promised! Not paying him was certainly a severe penalty, but we felt it was appropriate considering the problems he had caused us.

Don't expect clients to subsidize your failings. If you want to get paid, do what you promise. You're supposed to solve a client's problems, not add to them.

Corporate Games

We mentioned earlier how we were told by one client that our assignment really was to criticize their existing manuals, even though our accepted proposal just said we were to review existing manuals. Instead, we were directed to deliver a negative judgment. And in this chapter we explored how it appeared that our real task at one software company was to be the scapegoat for a project that was badly off-track. Eventually you'll find yourself in a situation where what you proposed to do, and what you thought you were supposed to do, is different from what the client really wants you to do. You'll seldom be told explicitly what the client really wants in these cases, especially if your role is to be the fall guy.

Even as an independent contractor, you can't completely escape corporate game playing. However, you have options that permanent employees don't. Since your relationship with the client is governed by a contract, you can rightfully refuse any task not covered by the contract. Be especially careful of situations where the client wants you to deliver a "directed judgment" as if it were your own independent conclusion. Your name will be attached to any such judgment, and you will have to live with the consequences of any conclusion made in your name. In the case of the manuals we were supposed to evaluate, we had already concluded they were bad and had no trouble delivering the verdict the client wanted. (However, we made no effort to solicit more assignments from that client—we felt there would be a time in the future when we would be expected to reach a conclusion we didn't agree with.) Other independent contractors have told us of cases where they have been "expected" to deliver reports recommending someone be fired, a whole department shut down, production be shifted to a new plant, etc. The client in these situations doesn't want independent judgment and conclusions; they have already reached a decision but want an "independent" seal of approval.

The signs you're being set up as a scapegoat are more subtle. Clues include an inability to get a response out of the company to your query, getting inadequate or incomplete information from the company, being unable to reach people within the company, and a steady pattern of missed deadlines and failure to perform their obligations as specified in your proposal. During our SuperTek experience, we wondered whether what we were experiencing was scapegoating or simple incompetence. Your best defenses are to document your dealings with the client, including firm letters when things start going wrong in a big way. And be willing to walk away from the project.

Your role is to perform the services specified in your proposal and nothing else. Don't let your clients drag you into their office politics and games.

Final Pearls Of Wisdom

The guidelines we've provided in the pages of this book aren't cast in stone, but they are based on our experience as both provider and consumer of the services of independent contractors. You'll make your own rules as you gain experience and confidence in your chosen field.

The next step is up to you. Remember, it may never seem like exactly the "right" time to go out on your own—you will always face a certain amount of doubt and uncertainty. But the rewards are real. Almost everyone we know who has taken the plunge says that they would never go back to working for a corporation. We certainly feel that way. If you do decide to make the break, we wish you the best of luck!

Index